High-Performance
AUTOMOTIVE
Cooling Systems

Covers Performance Street as well as Racing Systems

John F. Kershaw, EdD, PhD

CarTech®

CarTech®

CarTech®, Inc.
838 Lake Street South
Forest Lake, MN 55025
Phone: 651-277-1200 or 800-551-4754
Fax: 651-277-1203
www.cartechbooks.com

© 2019 by John F. Kershaw

Edit by Bob Wilson
Layout by Connie DeFlorin

ISBN 978-1-61325-504-9
Item No. SA462

Library of Congress Cataloging-in-Publication Data

Names: Kershaw, John F., author.
Title: High-performance automotive cooling systems / John F. Kershaw.
Description: Forest Lake, MN : CarTech, Inc., [2019]
Identifiers: LCCN 2018057158 | ISBN 9781613255049
Subjects: LCSH: Automobiles–Motors–Cooling systems. | Motor
  vehicles–Motors–Cooling systems.
Classification: LCC TL214.R3 K47 2019 | DDC 629.25/6–dc23
LC record available at https://lccn.loc.gov/2018057158

Written, edited, and designed in the U.S.A.
Printed in China
10 9 8 7 6 5 4 3 2 1

DISTRIBUTION BY:

Europe
PGUK
63 Hatton Garden
London EC1N 8LE, England
Phone: 020 7061 1980 • Fax: 020 7242 3725
www.pguk.co.uk

Australia
Renniks Publications Ltd.
3/37-39 Green Street
Banksmeadow, NSW 2109, Australia
Phone: 2 9695 7055 • Fax: 2 9695 7355
www.renniks.com

Canada
Login Canada
300 Saulteaux Crescent
Winnipeg, MB, R3J 3T2 Canada
Phone: 800 665 1148 • Fax: 800 665 0103
www.lb.ca

CONTENTS

I want to thank my wife, Joan, for her continued support in all of my projects.

I need to acknowledge my good friend, colleague, and noted author Jim Halderman, who has generously provided photographs, information, and support for this project along with many other projects. Jim introduced me to the idea of writing a book for CarTech.

I also want to thank Michael Harding of Champion Cooling Systems, Don Meziere of Meziere Enterprises, Emillia Teta of Evans Waterless Coolant, James Chartres of Kanga Motorsports, and Mike Murray of Eastwood Radiators along with assistance from BeCool Performance and Derale Performance for their generous technical information support and use of their copyrighted photographs.

PREFACE

I wrote this book to guide readers in selecting the best engine cooling system components and their placement for modified vehicles driven on the road or the track. Engine overheating issues often arise when older modified vehicles have high-displacement engines installed for street and track use. These older vehicles had low horsepower engines that were generally smaller in size than the engines that are replacing them. Often, this has led modified vehicles to have problems with low-speed cooling due to incorrect radiator and cooling fan selection. Others overheat at 65 mph in the middle of the day due to the wrong airflow across the radiator or an inadequate water pump. The correct placement of the right components is critical to cooling system efficiency.

To run correctly, the internal combustion engine cooling system must accomplish a harmonizing act. The system needs to remove enough heat to prevent overheating and provide efficient performance while also keeping the engine operating temperature in the 180 to 210°F range. To achieve and maintain optimum temperature range, an efficient engine cooling system needs the correct radiator and fan combination. It also needs the right water pump speed and coolant flow between the engine and radiator to keep the coolant in the radiator long enough to be cooled.

Typically, when an engine overheats or runs too cool, it is due to making one of several mistakes. These mistakes include using water as coolant, installing the wrong radiator, not using an anti-collapse spring in the radiator hose, installing a too-fast fan, not using a cooling fan or having too many cooling fans, installing the wrong radiator cap, using improper fan spacing and shrouding, or using inexpensive components. This book will hopefully dispel some of the common cooling systems myths and misconceptions that lead to incorrect practices and component selection by covering these mistakes.

This book will also cover thermodynamics, heat transfer, components and system operation; how to make your cooling system more efficient; and do-it-yourself service. It will provide information on how to select cooling system components for radiators, coolant, cooling fans, shrouds, thermostats, and more. Let's get started learning how to build a cooling system for your high-performance modified street or track vehicle.

ENGINE COOLING SYSTEM BASIC OPERATION

Automotive engine cooling systems have several functions. They remove excess heat from the engine. They help a cold engine quickly reach operating temperature and then maintain constant engine operating temperature. They also provide heat for the passenger compartment for street vehicles.

Understanding the basic operation of automotive cooling systems begins with a discussion on thermodynamics. Thermodynamics primarily deals with the conversion of one form of energy to another (heat that creates motion) along with the systems used to affect these conversions. The various concepts and laws that describe the movement of heat are called the laws of thermodynamics. These laws further explain the concepts of temperature and heat transfer, which are also important for understanding how a cooling system functions.

Thermodynamics

The main thermodynamic concept is energy, which is the ability to do work. Thermodynamics also deals with heat and temperature and their relation to energy and work. For example, energy can move from one object to another due to the difference in the objects' temperatures. When heat acts or moves the object through a distance, work is done.

We are discussing thermodynamics because heat is generated

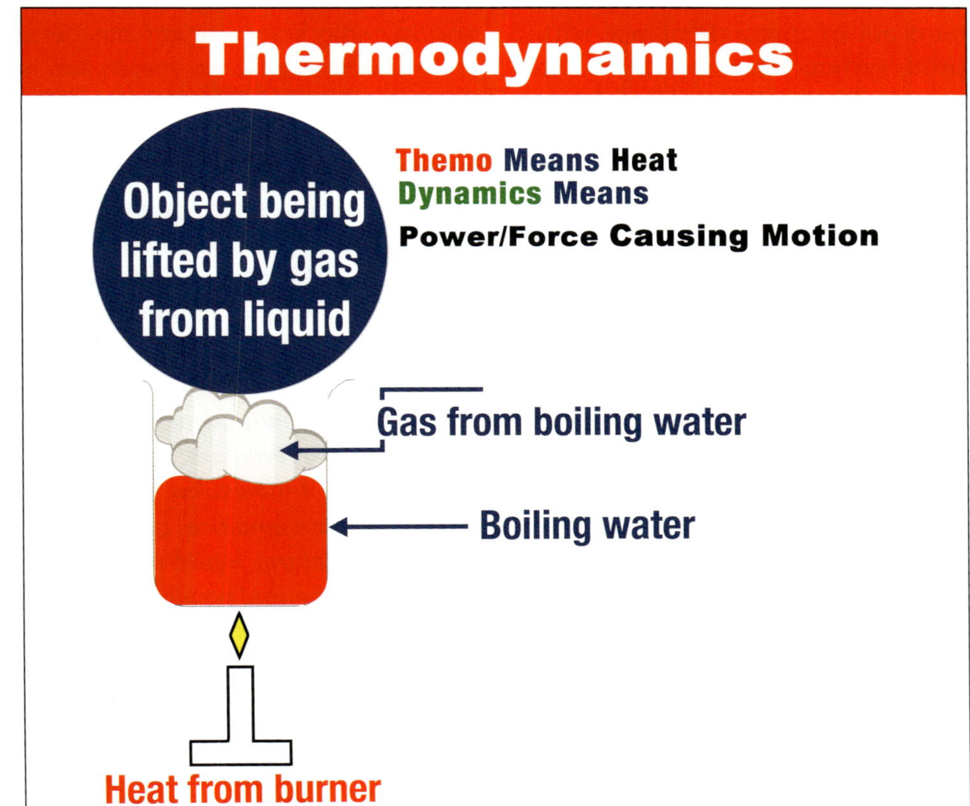

Thermodynamics is simply heat that is used to generate power and cause motion. It is an increase in internal energy of a closed system that is equal to the total of the energy added to the system. If the energy entering the system is supplied as heat and if energy leaves the system as work, the heat is accounted for as positive and the work as negative.

Systems and Surroundings

In thermodynamics, there are systems and surroundings. The system is composed of particles in balance and the quantity of matter or stuff under consideration. Everything else is called the surroundings.

There are open and closed systems. In a closed system, there is no interchange of matter between systems and surroundings. In an open system, there is an interchange, which is called a process. Any process or series of process where the system returns to its original condition is called a cycle. An internal combustion engine cooling system can be closed, open, or a combination of the two. ∎

in an internal combustion engine, and it needs to be removed by the engine cooling system. By understanding thermodynamics, anyone can design a cooling system that removes enough heat for efficient engine operation, which results in no overheating.

Thermodynamics developed in 1824 out of a desire to increase the efficiency of early steam engines when French physicist Nicolas Léonard Sadi Carnot wanted to help France with the Napoleonic Wars. In 1894, the definition of thermodynamics was stated as: "Thermo-dynamics is the subject of the relation of heat to forces acting between contiguous parts of bodies, and the relation of heat to electrical agency."

The early application of thermodynamics was later applied to mechanical heat engines. It was then applied to chemical reactions. Other forms of thermodynamics have evolved in the decades since.

Thermodynamic Laws

The laws of thermodynamics explain the movement of heat. There are four laws of thermodynamics: the Zeroth law, first law, second law, and third law. We will examine the three that apply to engine cooling system design.

Zeroth Law of Thermodynamics

The Zeroth law states: If two systems are in thermal balance with a third system, they are in thermal balance with each other.

This law helps define the concept of temperature. Temperature is the coldness or hotness of something solid, liquid, or gas. It is the measure of the speed of the molecule vibration of that thing. An increase in temperature indicates that the speed of the molecules has increased. A drop in temperature means a decrease in molecular speed. When heat is removed from an item, it becomes cold with a lower temperature. Heat always flows from the warmer object to the colder object.

Temperature is not a measure of the total quantity of heat; instead, it is a measure of the degree of heat that something possesses. Temperature tells you whether something has gained or lost heat. A thermometer is used to measure temperature, and in today's auto repair business, an infrared digital thermometer is used.

The First Law of Thermodynamics

The first law of thermodynamics states that heat is a form of energy. It is subject to the principles of conservation, which states energy can neither be created nor destroyed, only

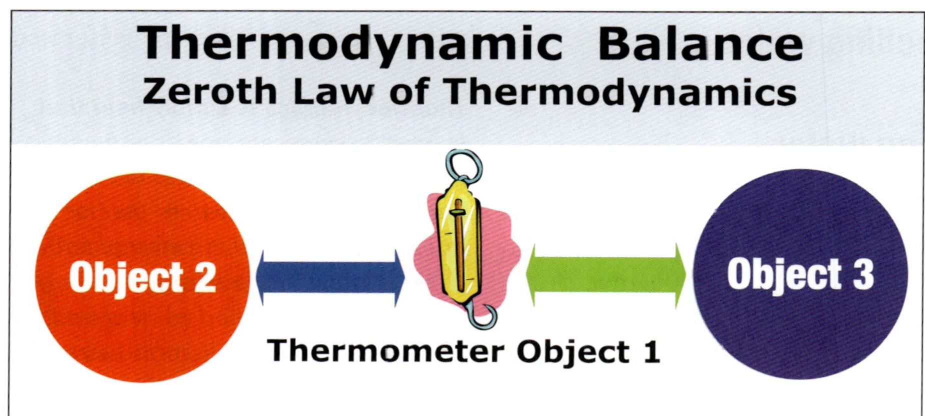

The Zeroth law of thermodynamics deals with the concept of temperature, or the speed of an item's molecules. It allows the existence of an empirical parameter (the temperature), which means that systems in thermal equilibrium with each other have the same temperature. A particular physical body, for example a mass of gas, can match temperatures of other bodies, but that does not mean the temperature is a quantity that can be measured on a scale of real numbers.

An infrared digital handheld noncontact thermometer is used to measure temperature. These are available online or in retail stores for about $45. It is pointed at whatever needs to be measured and the temperature reading appears on the digital screen. (Photo Courtesy James Halderman)

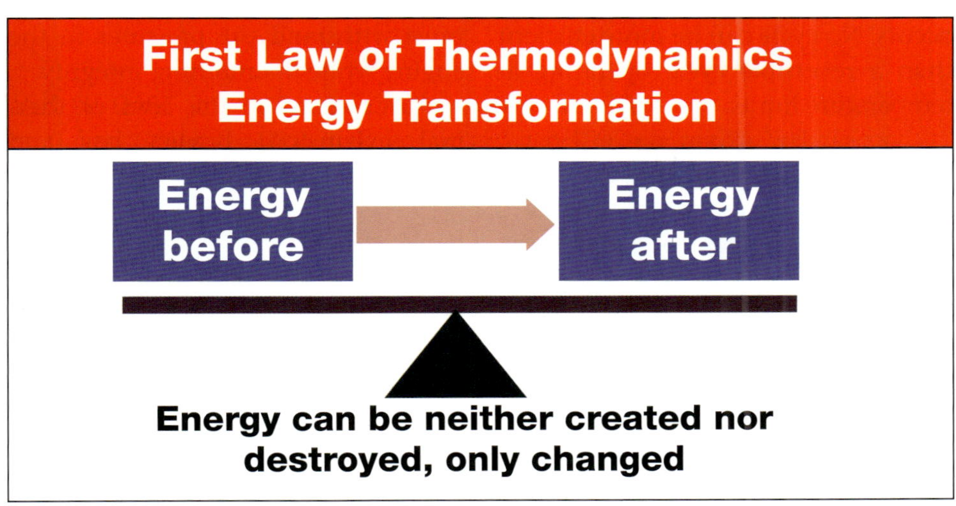

The first law of thermodynamics states that when energy passes (as work, as heat, or with matter) into or out from a system, the system's internal energy changes in accord with the law of conservation of energy.

changed. This means that a perpetual motion machine of the first kind (a machine that produces work with no energy input) is not possible.

The Second Law of Thermodynamics

The second law of thermodynamics says that the entropy of a closed system must either increase or stay the same. *Entropy* has a very precise definition in science that is not commonly known by most people. It can be roughly related to the level of disorder or the loss of information or the amount of useless energy; that is, energy that cannot be used to perform work. For example, a tool drawer system that is disorganized has a higher entropy than an ordered tool system with tools neatly arranged. Similarly, a state in which information decreases or the amount of useless energy increases can be said to be a state in which entropy is increasing.

The second law also says that heat transfer always flows from the warmer object to the colder object. When a block of ice melts, it is absorbing heat flowing into it. Engine coolant is cold, so heat from hot engine combustion flows to the colder coolant. Heat cannot sponta-neously flow from a colder location to a hotter location.

The second law is an observation of the fact that differences in temperature, pressure, and chemical potential tend to even out over time in a physical system that is isolated from the outside world. The conversion of heat to work is limited by the temperature at which the conversion

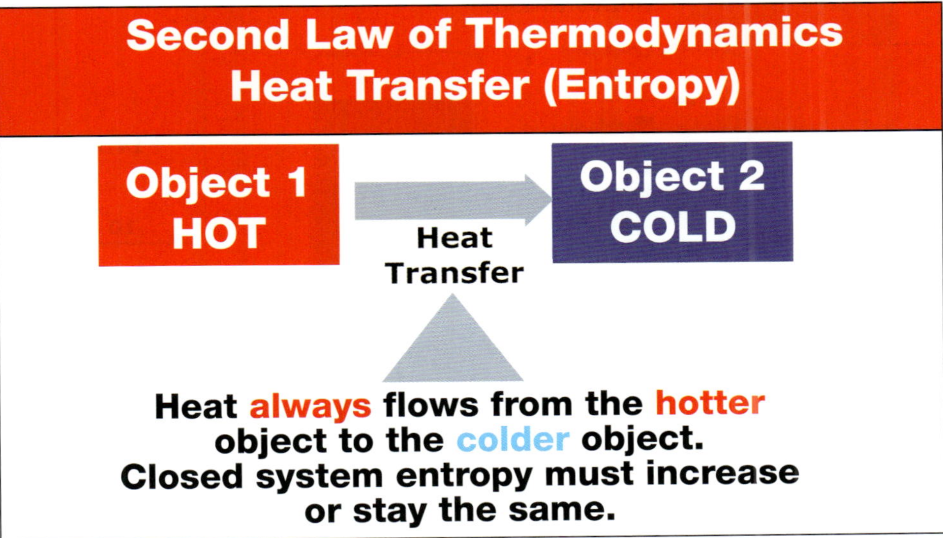

The second law of thermodynamics states the sum of the entropies (disorder levels) of the interacting thermodynamic systems increases. This is to say that a perpetual motion machine of the second kind (a machine that spontaneously converts thermal energy into mechanical work) is impossible.

occurs. No cycle can be more efficient than a reversible cycle operating at temperature limits.

Thermal Expansion

Another aspect of thermodynamics involves thermal expansion. Thermal expansion is the dimensional changes exhibited by solids, liquids, and gases during changes in temperature at constant pressure. As solids, liquids, and gases heat up, they expand, taking up more room. These changes affect the amount of pressure in a vessel.

The solid engine block must constantly and efficiently deal with different types of liquids and gases. Engine coolant has to withstand drastic temperature changes, and liquid expands and contracts based on these differences. Throughout the expanding and contracting events, the coolant must withstand the pressures in the engine while maintaining integrity.

Let's examine the laws of thermal expansion: Boyle's law and Gay-Lussac's law.

Boyle's Law

Boyle's law states that the pressure exerted by a mass of a perfect gas is inversely proportional to its volume if it remains at a constant temperature and the volume does not change. In other words, at a constant temperature, the amount of gas depends on the volume in the vessel holding it (high pressure/low volume). If volume increases, the pressure decreases, and vice versa.

Boyle's law applies to compression in a spark ignition engine as well. When the piston rises on compression, cylinder volume is reduced and the pressure increases. This exerts pressure on the cylinder walls when the same number of molecules now have less space to move.

Gay-Lussac's Law

Gay-Lussac's law, or the pressure law, was discovered by Joseph Louis Gay-Lussac in 1809. This law states for a given mass and constant volume of an ideal gas, the pressure exerted on the sides of its container is directly proportional to its absolute temperature. So, if a gas's temperature increases, the particles move faster and create more pressure.

Pressure

Pressure must also be included in a discussion of thermodynamics. Atmospheric pressure is the force that is exerted per unit area by an atmospheric column (the entire body of air above an area). It is measured by a barometer and expressed in several

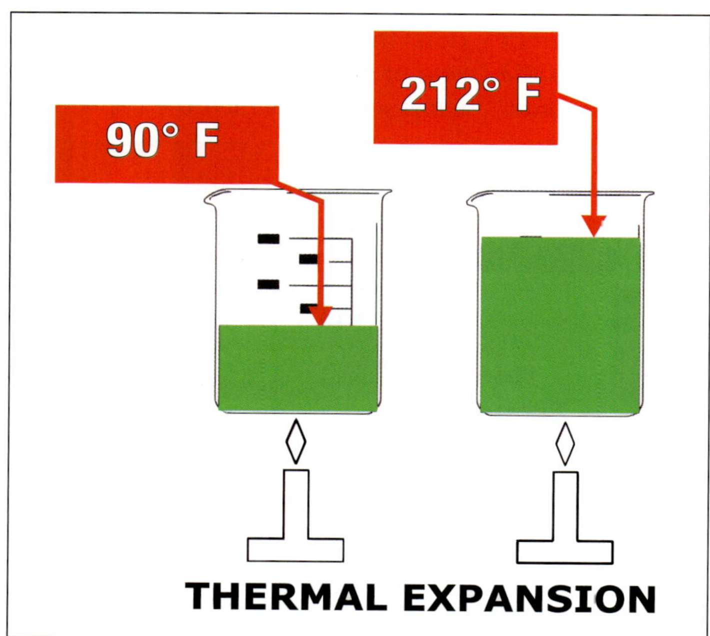

Thermal expansion is the expansion of matter (solid, liquid, or gas). Shown here are two beakers of engine coolant being heated by Bunsen burners. The liquid coolant in the beaker on the right expanded when heated to the boiling point of water (212°F).

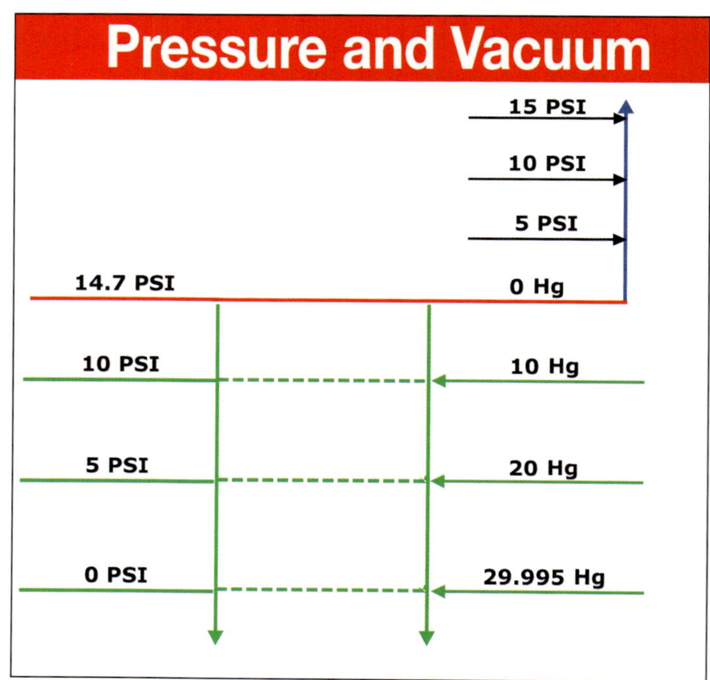

This diagram shows atmospheric pressure and vacuum. The red center line is at atmospheric pressure of 14.7 psi (sea level) with the base line of 0 inches of Mercury (Hg). When the pressure decreases, the vacuum increases.

different systems of units: inches (or millimeters) of mercury (Hg), pounds per square inch (psi), dynes per square centimeter, millibars (mb), standard atmospheres, or kilopascals (kPa). For our purposes, we will use pounds per square inch (psi).

Atmospheric pressure at sea level is 14.7 psi and decreases as elevation increases. At 5,000 feet (1,524 m) above sea level, a 1 square inch column of air from the earth's surface to the outer edge of the atmosphere is 5,000 feet (1,524 m) shorter than the same column at sea level. Therefore, the weight of this column of air is less at 5,000 feet (1,524 m) elevation than at sea level. As altitude continues to increase, atmospheric pressure continues to decrease.

Vacuum

Vacuum is the absence of pressure or just low pressure. When a space contains a vacuum, it contains a smaller quantity compared with the amount of air the same space is capable of containing (as dictated by atmospheric pressure). Vacuum could be measured in psi, but inches of mercury (Hg) is more commonly used. A complete vacuum or absence of pressure is equal to 29.995 Hg (101.06 kPa).

Atmospheric pressure and vacuum are used in all automotive systems. For example, atmospheric pressure is available outside the engine air intake. When a piston moves downward with an intake valve open, a vacuum is created in the cylinder above the piston. The air moves rapidly from the high pressure outside the air intake to the lower pressure in the cylinder.

Heat Transfer

Now that we've discussed the relationship between temperature, pressure, and volume, it is time to examine how heat affects energy. Heat is thermal energy and cannot be destroyed, only transferred. It always moves from a hotter object to a colder object, as we saw in the second law of thermodynamics.

Heat is transferred in three ways: conduction, convection, and radiation.

Conduction

Conduction is the transfer of heat from molecule to molecule through solids and fluids in intimate contact at rest. It is heat transfer from one solid to another. No displacement of the heated body takes place during conduction. The heat travels through a rod via conduction from molecule to molecule until the end you are holding is near the temperature of the end in the fire. The action of a solid to conduct heat is called conductivity.

Convection

Convection is heat transfer by the molecular motion in the heated substance itself. It only takes place in liquids and gases. There are two forms of convection heat transfer. Natural convection is when the fluid motion is caused by different densities in a gravitational field. Forced convection is the method of heat transfer between a fluid and a solid surface in relative motion, when the motion is caused by forces other than gravity.

Most of the heat in a combustion engine flows between the coolant (working fluid) and the engine parts, and it is transferred via forced convection. Heat transfers by circulation though fluids, such as coolant (air in some engines), in motion between the fluid and a solid surface in motion, such as a piston. It is heat from combustion that is transferred to the cylinder wall, transfers to the coolant, and is carried away at the radiator. This form of heat transfer includes conduction as well as fluid motion.

Radiation

Radiation is the transfer of heat through space. This takes place in a

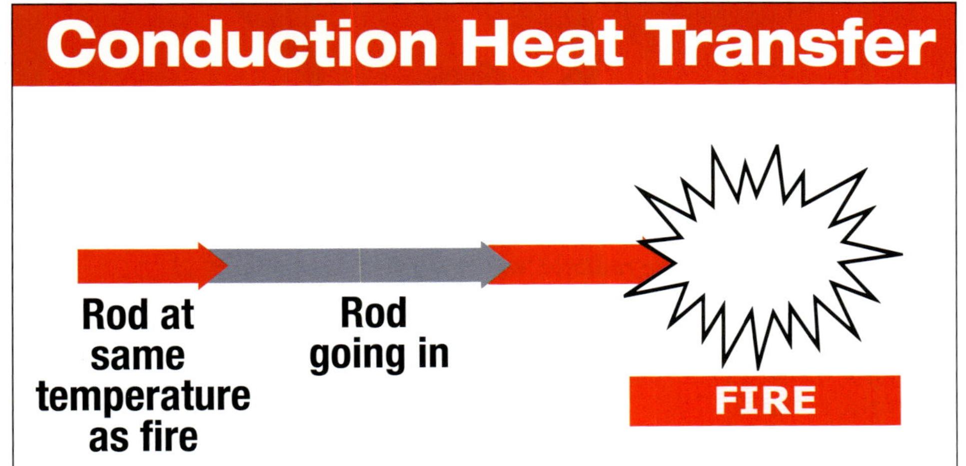

The textbook definition of conduction in the sense of heat transfer is transmission through or by means of a conductor. Conduction is heat transfer from molecule to molecule through solids in contact. A rod going into a fire transfers heat from the fire to the other end of the rod, and if you are holding it, ouch.

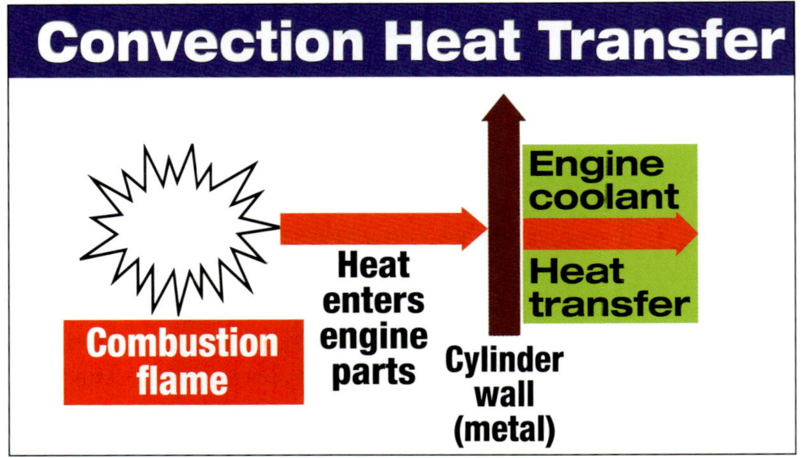

The textbook definition of convection is the action or process of conveying movement in a gas or liquid in which the warmer parts move up and the cooler parts move down. This graphic shows the heat transfer by molecular movement in the heated substance itself, or heat transfer by circulation through the engine coolant.

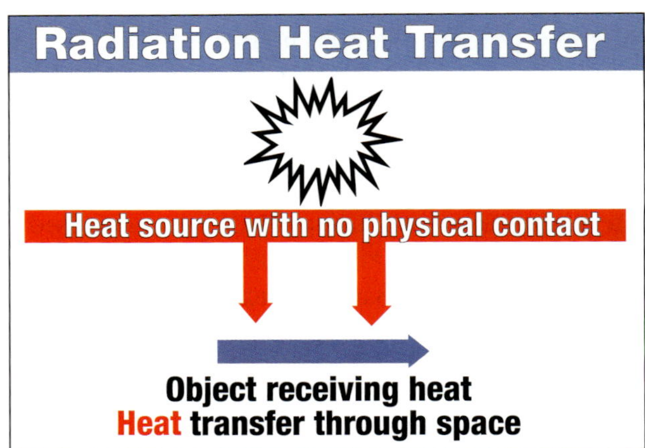

Radiation is the process of emitting radiant energy in the form of waves or particles. It is the combined processes of emission, transmission, and absorption of radiant energy. The heat source has no physical contact with the object receiving the heat and the heat is transferred to this object.

vacuum (absence of pressure or very low pressure) or through solids and fluids that are transparent to wavelengths in the visible and infrared range. A small fraction of the heat transferred to the engine cylinder walls from the hot combustion gases transfer via radiation. The radiator in an engine cooling system is the main heat exchanger.

Cooling System Operation

In a basic cooling system, coolant is drawn from the radiator by the water pump, and this pump pushes the coolant into both sides of the cylinder block. Coolant flows around the cylinders and up into the cylinder heads, where it circulates around the exhaust passages and fire deck areas of the cylinder head. The coolant then flows out of the front into the thermostat housing. This housing stops the main flow when the engine temperature is below the opening temperature of the thermostat. It then directs a small portion back to the pump through a bypass hose or

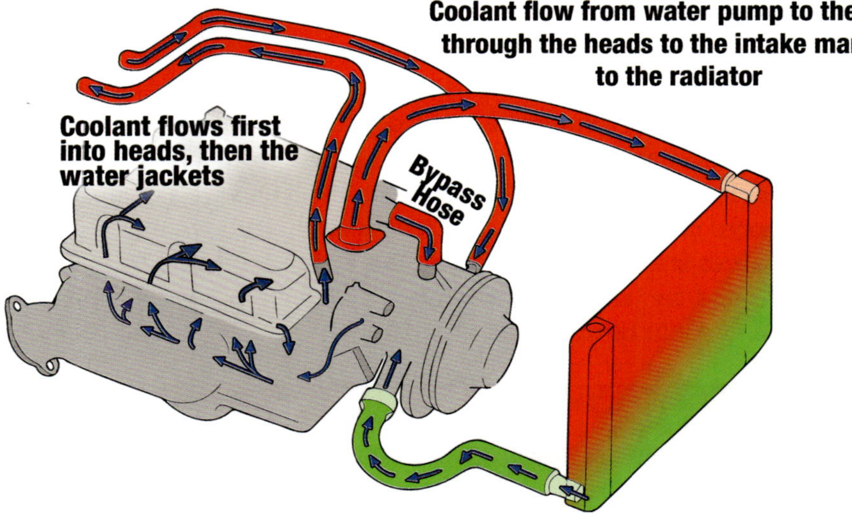

Engine coolant flow starts at the water pump with cooled coolant pumped into the cylinder block. It then goes through the heads, which is the hottest part of the engine. The coolant carries the heat of combustion through the intake manifold and thermostat to the radiator, where the heat is removed.

passage. When the thermostat is open, the coolant flows back into the top of the radiator.

Reverse-Flow Cooling System

From 1992 until 1996, General Motors used a reverse-flow cooling system in the Chevrolet Corvette LT1 Engine. In this system, coolant flows to cool the heads before the cylinder block. There are some import car manufacturers that use this type of system, but all of the domestic original equipment manufacturers (OEMs) use the traditional flow system with the cylinder heads receiving the coolant last before returning to the radiator for cooling.

In the traditional system, the water pump draws coolant from the radiator and the coolant circulates through the cylinder heads first via the water pump and then into the cooling jackets in the engine block. Coolant is then directed back to the radiator where it's cooled. In the reverse-flow cooling system, the water pump draws coolant from the radiator and the coolant passes through the thermostat on the inlet side of the pump. Vapor is vented off (if there is any) through the air bleed pipe, and the coolant travels down through the bores and into the block. Once the coolant leaves the engine block, it returns to the water pump, where the coolant travels through a cast internal passage to the upper radiator hose back to the radiator for cooling.

The system directs some coolant through hoses to the heater core. A sealed recovery or expansion reser-

Reverse-Flow System Components

The reverse-flow cooling system consists of these components:

- Air bleed venting circuit
- Coolant recovery reservoir or expansion tank
- Cooling fans
- Cylinder heads that are unique for LT1 block
- Hoses
- Radiator
- Surge tank
- Thermostat
- Water pump
- Water pump gear driveshaft

voir connects to the radiator surge tank in order to retrieve the coolant displaced by expansion. As the coolant cools and contracts, vacuum draws the coolant back into the radiator surge tank. The radiator surge tank provides a coolant fill point and a central cooling system air bleed location. A sensor can be used in the radiator surge tank to show the cool-

ant level. When the coolant in the system falls below the recommended level, a warning lamp in the instrument panel turns on.

Cooling System Components

Coolant circulates through an engine and absorbs excess heat through the convection heat transfer

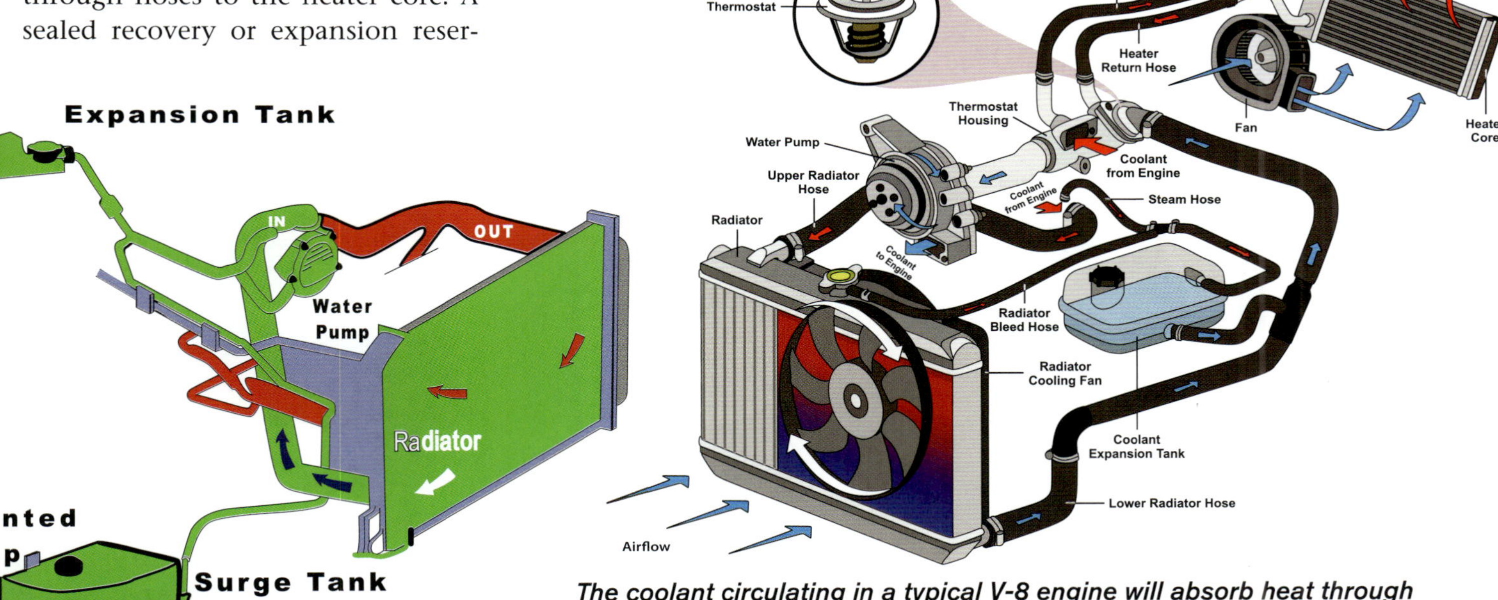

In a 1992–1996 Chevrolet Corvette LT1 engine reverse-flow cooling system, coolant flows to cool the heads before the cylinder block.

The coolant circulating in a typical V-8 engine will absorb heat through the convection process. The flow in this illustration is from the hot side to the cold side of the system. Hot coolant returning with the heat of combustion in the coolant goes into the top or inlet side of the radiator when the heat is transferred to the air moving through the radiator cooling fins over the cooling tubes. It is being sucked out of the radiator at the outlet or bottom by the water pump, where it returns to the engine to start the cooling process over again.

process. A combustion engine cooling system can be closed or open. In a closed system, the engine coolant has no contact with outside air. In an open system, the coolant has contact with the outside atmosphere.

Open systems are things of the past with the exception of marine applications. All current systems are a combination of the two, where the coolant circulating through the engine has no contact with the outside air, but at the radiator pressure cap there is contact through the vacuum or air valve, which allows the entrance of atmospheric pressure.

When designing a new engine cooling system for a street rod or track vehicle, the selection of cooling system components is very important for engine life and performance. Consider the following components carefully.

Water Pump

Early engines relied on thermosiphon cooling. In this system, hot coolant left the top of the engine block and passed to the radiator, where it was cooled before returning to the bottom of the engine. Circulation was powered by convection alone with no moving parts.

Modern liquid-cooled engines usually have a circulation pump, which in automotive lingo is called a water pump. The water pump is located between the engine block and the cooling fan. It is driven by the crankshaft through belts and pulleys or directly using a gear. It also has a fan-shaped impeller set in a round chamber (called a volute) with curved inlet and outlet passages. In addition to the housing, the pump also has a coolant inlet and outlet.

The water pump uses centrifugal force to circulate the coolant. To do this, the impeller rotates, forcing the coolant through the engine. The inlet is attached to the bottom of the radiator, where cooled coolant is sucked out of the radiator via the lower radiator hose and flow out the outlet into the engine via the upper radiator hose.

The typical discharge rate is 70 gallons per minute (gpm) with a cooling system capacity of 6 gallons with 3 gallons in the engine. An engine on an engine dynamometer will show the actual amount of coolant flowing through the cooling system.

Water Distributing Tube

Some engines of the past, such as the Ford flathead V-8, used a water distributing tube inserted in the water jacket near the exhaust valve. The front end of the tube received water directly from the water pump. Holes in this tube had discharge jets that provided extra cooling for the exhaust valve seats and stems.

Radiator

Radiators may be manufactured from a range of electrochemically incompatible metals (aluminum, cast iron, copper, brass, solder). Most internal combustion engines are liquid cooled. These engines use a mixture of water and chemicals, such as antifreeze and rust inhibitors called engine coolant, run through a heat exchanger called a radiator, which is cooled by airflow.

The water pump uses centrifugal force to circulate the coolant. It consists of a fan-shaped impeller set in a round chamber called a volute with curved inlet and outlet passages. (Photo Courtesy Jim Halderman)

This high-flow water pump runs in excess of 6,000 rpm and can deliver up to 120 gpm of engine coolant. (Photo Courtesy Meziere Performance)

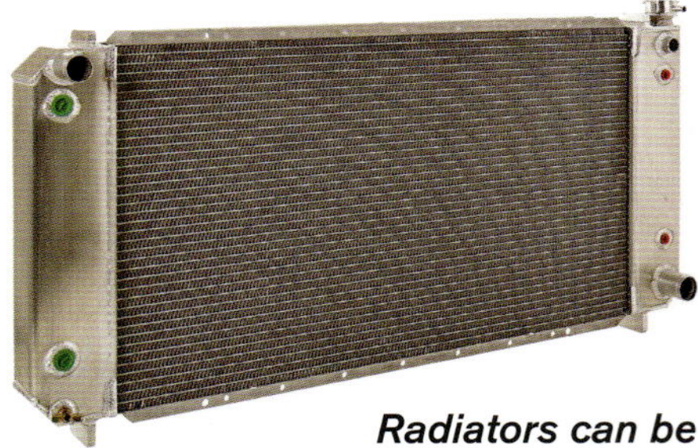

Radiators can be downflow or crossflow type, such as the Dual Cooler tank (shown). They can also be designed with a single, dual, or triple pass of coolant through the system. The radiator system may also use a separate expansion tank outside the radiator, which is partially filled with coolant and is connected to the radiator pressure cap. (Photo Courtesy Derale Performance)

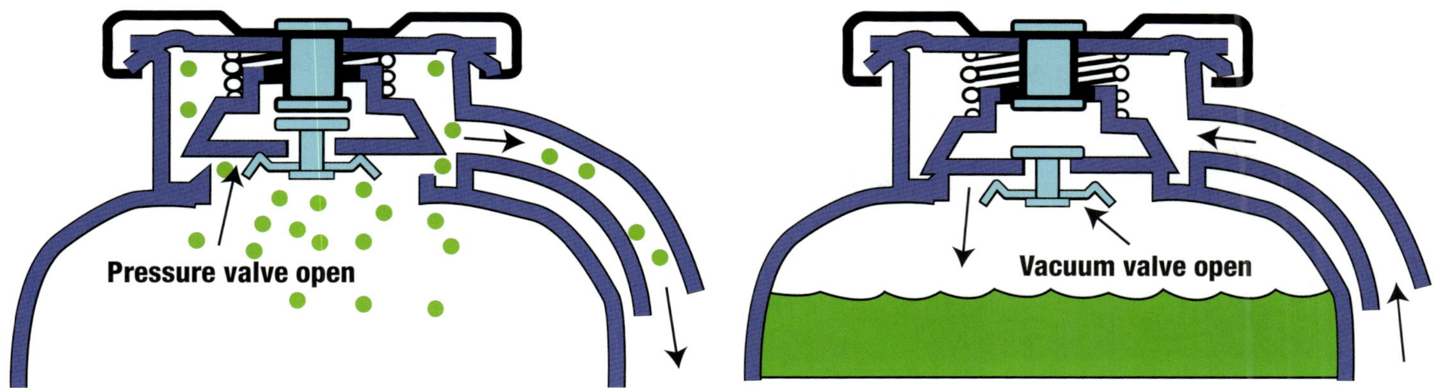

The radiator pressure cap uses a pressure valve (left) controlled by a spring that holds pressure in the system until it reaches a specified pressure and then opens. It also contains a vacuum valve (right). As temperatures drop and the coolant contracts, a vacuum is created in the engine's cooling system. On older open cooling systems without tanks, the vacuum valve opens to prevent the end tanks from being crushed or imploded. On a closed system with an overflow/ recovery reservoir or expansion tank, the vacuum valve opens and allows coolant to flow from the overflow tank back into the radiator. As pressure inside the system drops, outside air pressure helps the coolant flow in.

The radiator holds a large column of coolant in close contact with the flowing air. This allows the radiator to transfer heat via convection from the coolant to the outside air. Radiators can be downflow or crossflow type, and they can have a single, dual, or triple pass of coolant through the radiator design.

The radiator system may also use a separate expansion tank outside the radiator, which is partially filled with coolant and is connected to the radiator pressure cap. Coolant expands as it heats up and sends part of the coolant into the expansion tank.

Radiator Pressure Cap

A radiator cap allows an engine's coolant (water and antifreeze) to expand and contract without allowing air to enter the cooling system. An upper seal protects the system at all times. After the engine warms and system pressure reaches the cap's rated pressure, a pressure spring compresses and pressurized coolant flows into a reservoir or a coolant overflow tank. This allows for expansion of the heated fluid.

The boiling point of water is always 212°F under standard atmospheric pressure (14.7 psi). If the coolant in a closed system is under pressure greater than 14.7 psi, the boiling point of the coolant will be higher. This occurs because the coolant molecules are compressed by the pressure and will have to vibrate more for the temperature to increase. For every pound of spring pressure, the boiling point is increased by 3 degrees. This means that a cap rated at 15 pounds will increase the boiling point in a system by 45°F with a boiling point of 257°F (212 + 45).

The radiator pressure cap also has a vacuum valve that allows the coolant to flow back into the radiator as the engine cools. As temperatures drop and the coolant contracts, a vacuum is created in the engine's cooling system. The vacuum valve opens and allows coolant to flow from the pressure cap and either onto the ground (early cooling systems) or back to the radiator (later systems).

As the cap wears, the spring weakens and excessive coolant is allowed to flow into the reservoir tank. This overflow will result in a loss of coolant at the seal between the cap and the radiator as well as the overflow tank. If the system passes a coolant system pressure test with no leaks in the engine or passenger compartments, suspect a faulty radiator cap.

High-performance radiator pressure caps range from 19 to 32 psi. If needed, engines with higher operating temperatures can be designed with OEM parts.

Overflow, Recovery, Reservoir Tanks

When coolant expands from engine temperatures rising, it typically creates steam, which needs to be vented in order to protect the cooling system. Vehicles from the 1950s, 1960s, and 1970s had a hose attached to the radiator's filler neck below the cap that vented to the atmosphere, dripping on the ground just below the radiator. The steam was able to escape; however, there was another issue. As steam expands so does the pressure, and often the vented cap also expelled coolant through the hose and onto the ground. This system was good for engine cooling but not good for the environment.

This vintage reservoir tank was found on a C3 (third-generation Corvette built from 1968 to 1982), which was the old Stingray version.

In the late 1970s, environmental control regulations required adding a reservoir to that vent tube. This allowed the steam to be expelled and captured the coolant that came with the steam. Capturing that expelled coolant meant that it could then be recovered and reintroduced into the radiator, hence the name recovery or reservoir tank. It is also called an overflow tank. For this book, we will refer to them going forward as reservoir tanks.

Returning coolant to the radiator is possible because the reduced steam pressure allows the atmospheric pressure to push coolant from the reservoir tank back into the radiator through the vented radiator cap. This adds more coolant to your system and helps to keep the engine cooler.

Reservoir tanks use full cold/hot marks to indicate the ideal coolant level while an engine is cold (non-running) and hot (running). These tanks also have a vent. Expelled coolant will enter the tank from the bottom, and when the level rises it will expel through the vent tube. In contrast to the expansion tank, the reservoir tank featured a vented cap and was not required to be above the cylinder heads.

Catch Tanks

A catch tank also collects expelled coolant, but it will be drained from that tank at a later point. You should not confuse the reservoir tank with a catch tank. While both the reservoir tank and the catch tank hold excess coolant, the reservoir tank will automatically put the coolant back in the system, where the catch tank will hold the coolant until it can be emptied.

The reservoir tank will either be plumbed at the bottom of the tank or have a hose internally that runs to the bottom so coolant can be drawn back into the system. A catch tank generally uses plumbing into the top of the tank and does not have a hose that goes to the bottom.

Expansion Tanks

Expansion tanks (also called surge tanks) provide expansion of your cooling system, giving you up to a half gallon more coolant. They are designed so there is space in the tank for the coolant to expand, hence the name. If an expansion

Polymer or plastic tanks have full cold/hot marks that indicate where the coolant should be while cold (non-running) and hot (running). (Photo Courtesy Jim Halderman)

An expansion tank with a high-pressure cap provides an area for the expansion of the cooling system, as the name implies. They typically add about a half gallon of additional coolant to the system. (Photo Courtesy Jim Halderman)

Do not remove a pressure cap while the engine is hot. (Photo Courtesy Jim Halderman)

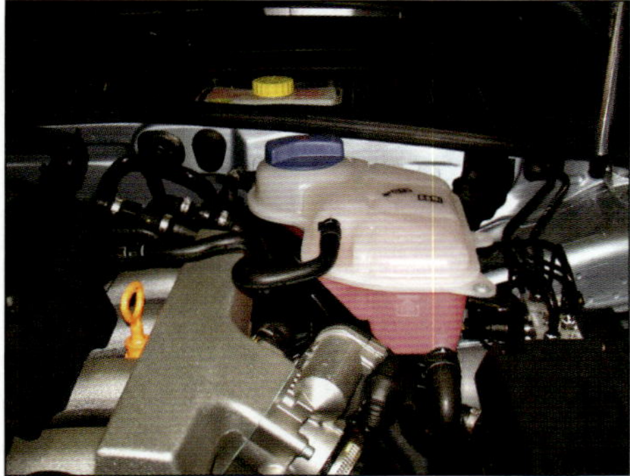

Coolant expands as it heats up and sends part of the coolant into the expansion tank. (Photo Courtesy Jim Halderman)

This late-model turbocharged Chevrolet Cruze uses an expansion tank with a 20 psi pressure cap sealing the cooling system. This system also has a bleed line coming out the top of the expansion tank that returns to the top of the cylinder head.

tank is overfilled, it will discharge coolant when the system is at operating temperature and can also be used as a system fill point.

When an expansion tank is used, the radiator doesn't require a pressure-relieving cap because the expansion tank cap does all of the things that a traditional radiator cap performed. The expansion tank needs to be located above the cylinder heads. That way the additional coolant stored in the expansion tank can flow back into the main cooling system when the engine cools down.

Expansion or surge tanks are usually sealed systems and not affected by atmospheric pressure. When the engine reaches operating temperature, the pressure valve in the cap closes and seals the system. As temperatures and pressures increase, the expansion tank allows the cooling system to expand without bleeding any off to the atmosphere.

With this type of system, the radiator's cap will be a non-vented; the expansion tank will have the vented cap in case the pressure exceeds the expansion tank's capacity. The expansion tank also has a connection to the cooling system through a bypass or a heater hose. Instead of just containing expanded coolant, the expansion tank is part of the cooling system and coolant circulates through the tank as it

The expansion tank for a Champion Cooling Systems' aftermarket expansion tank uses a vented cap, an upper hose that goes to the radiator filler neck below the cap, and a lower hose that connects to the heater core. (Photo Courtesy Champion Cooling Systems)

does through the radiator. When the system balances out, you should not have to add additional coolant.

Recovery Expansion Tank System

The coolant recovery tank system can also be an expansion tank with the radiator pressure cap on the radiator, but this tank is above the cylinder heads and connected to the radiator fill neck with the overflow tube at the bottom of the tank. This way, coolant can flow back and forth between the radiator and the reservoir.

This cooling system with a coolant recovery reservoir is still a closed system. When the pressure within the cooling system gets too high, the pressure valve in the pressure cap will open. This allows the expanded coolant to flow through the overflow tube and into the recovery reservoir. As the engine cools down, the temperature of the coolant drops and a vacuum is created in the cooling system. This opens the vacuum valve in the pressure cap, allowing some of the coolant in the reservoir to be siphoned back into the radiator.

Under normal operating conditions, no coolant is lost in this system. Although the coolant level in the recovery reservoir goes up and down, the radiator and cooling system are kept full. An advantage to using a coolant recovery reservoir is the elimination of almost all air bubbles from the cooling system. Coolant without bubbles absorbs heat much better than coolant with bubbles.

Thermostat

The cooling system thermostat is located in the coolant passage

The cooling system thermostat closes off a coolant passage between the cylinder head or intake manifold and the radiator inlet to control engine operating temperature. It is the temperature control valve for the cooling system. (Photo Courtesy Jim Halderman)

This 2016 Chevrolet Traverse coolant recovery tank system uses a plastic coolant recovery reservoir and overflow tube that is partially filled with coolant. It is connected to the radiator fill neck with the overflow tube at the bottom of the tank.

Cooling system thermostats use a wax pellet that expands when heated and contracts when cooled down. (Photo Courtesy Jim Halderman)

between the cylinder head or intake manifold and the radiator inlet at the top. Its purpose is to close off this passage when the engine is cold so that coolant circulation is restricted. This allows the engine to reach normal operating temperature more rapidly.

The thermostat is designed to open at a specific temperature, such as 180°F or 190°F. When the engine is cold, the coolant uses a bypass hose or passage (blocking thermostat) to circulate the coolant through the cylinder heads and engine block.

Most cooling system thermostats use a wax pellet that expands when heated and contracts when cooled down. This pellet is connected through a piston to a valve. When the pellet heats up, pressure is exerted against a rubber diaphragm that forces the thermostat valve open, allowing coolant flow. When the pellet cools down, the contraction of the pellet causes a spring to close the valve, cutting off coolant flow.

Cooling Fan

When a vehicle is moving, air is directed up and through the radiator so that the heated coolant can be cooled through the convection process. Cooling system fans are used to blow air through the radiator when the vehicle is stationary.

Some longitudinally mounted engines use mechanically driven cooling fans. Some of these mechanical fans use a thermostatically controlled viscous clutch but not always. This clutch is positioned at the hub of the fan, in the airflow coming through the radiator. This viscous clutch is much like the viscous coupling sometimes found in all-wheel-drive (AWD) vehicles.

Newer vehicles and all front-wheel-drive (FWD) vehicles use

The electric engine cooling fan (left) is supplementing a viscous mechanical cooling fan (right). In most systems, you would use either an electric cooling van or a viscous clutch mechanical fan, but rarely both. (Photo Courtesy Jim Halderman)

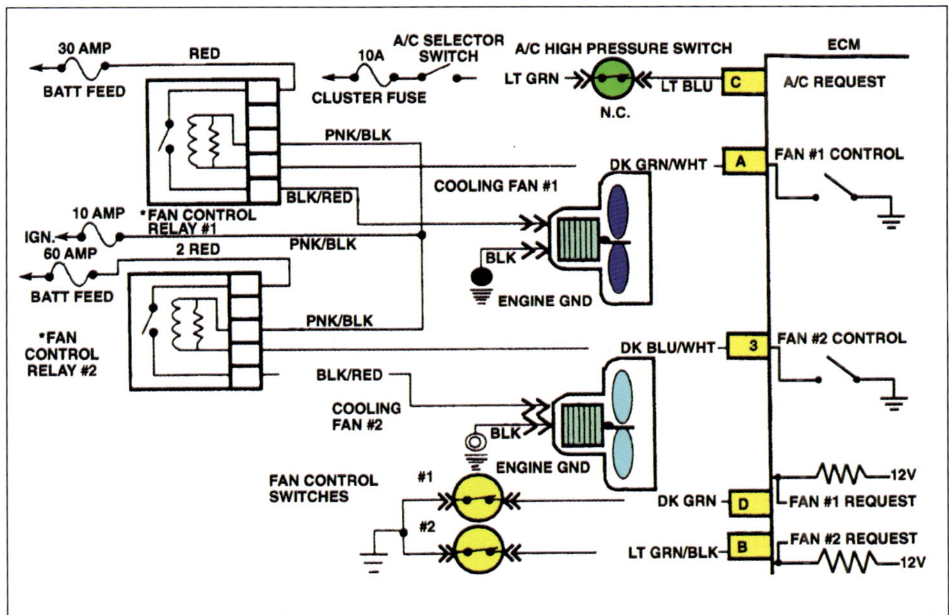

The electric cooling fan circuit is operated by the engine management computer, in this case an engine control module (ECM). Based upon input from fan #1 and #2 control switches, along with the air conditioner high-pressure switch, the ECM will ground fan control relay #1 or both fan control relay #1 and #2 to turn on the cooling fans.

electric fans. The fans are controlled either with a thermostatic switch or by the engine computer. They turn on when the temperature of the coolant goes above a specific temperature. The fan turns back off when the temperature drops below that point. The cooling fan has to be controlled so that it allows the engine to maintain a constant temperature.

FWD vehicles use electric fans because the engine is mounted transversely, meaning the output of the engine points toward the side of the car. For all vehicles built after 1996 that use Onboard Diagnostics Generation II (OBD II), the ECM will use the engine coolant temperature (ECT) sensor to signal the fan.

Temperature Switch/Sensor

Early vehicles and those built for street or track may use a temperature switch or ECT sensor to operate a temperature gauge or a dash warning light. These display the engine's temperature or alert the driver that the engine is overheating. In some cases, the sensor will turn on the electric cooling fan(s). Most all vehicles from about 1994 and all vehicles under 8,500 pounds gross vehicle weight after 1996 use an ECT sensor.

The ECT circuit is a variable ground that uses a voltage divider network. In the network, the voltage is divided between the sensor input and a sensor ground inside the computer (also known as the black box). The engine control module (ECM) or powertrain control module (PCM) provides a 5-volt reference signal to the ECT sensor.

When cold, the sensor provides high resistance, which the computer reads as high signal voltage. As the engine warms up, the thermistor sensor resistance becomes lower

and the signal voltage drops, so that is the difference. You do not need to know all of the inner workings of the computer, just that when you check voltage on the yellow wire of a cold engine, you should read high voltage at 3 to 5 volts. At normal operating temperature, the signal voltage should be around 0.47 to 1.45 volts.

The engine coolant temperature (ECT) sensor is screwed into a water jacket somewhere on the cylinder head. It is the temperature-sensing device for all vehicles that use Onboard Diagnostics Generation II (OBD II). A 2.2L GM Ecotec 4-cylinder engine is shown here.

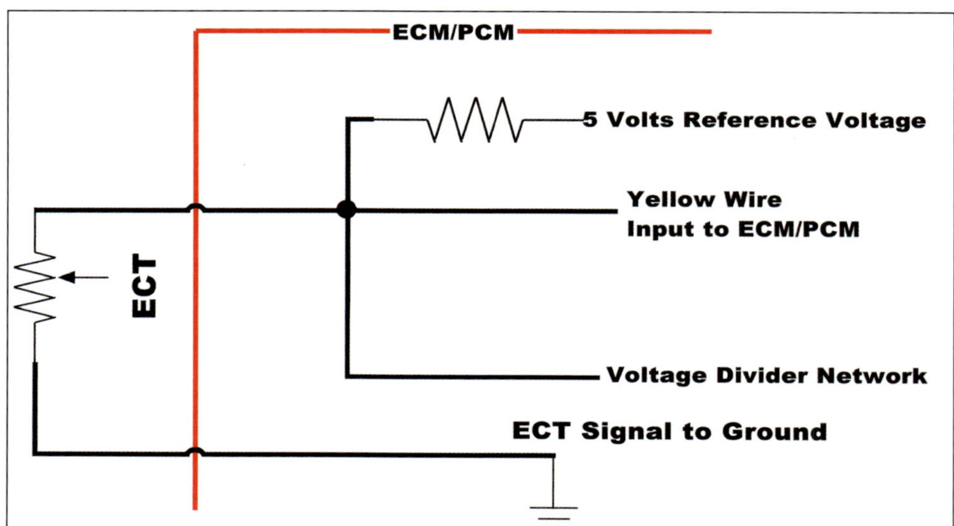

The ECT sensor provides the engine coolant temperature data to the engine management computer such as an engine control module (ECM). Problems in this circuit will turn on the malfunction indicator lamp (MIL) on OBD II built after 1996 or the check engine light on earlier systems. It will set a diagnostic trouble code (DTC).

COOLANT SELECTION

Coolant is one of the most important components of your high-performance cooling system. It should never be overlooked because it transfers heat from the engine to the radiator, protects the engine and the cooling system from rust and corrosion, and prevents freezing in cold climates.

Most automotive engines are water-cooled to remove waste heat. Although the term *water* actually means a coolant composed of antifreeze and water, not water alone. The automotive industry uses the term *engine coolant*, which covers its primary function of convective heat transfer for internal combustion engines.

Coolant Is Antifreeze and Water

All coolants are a mixture of antifreeze and water, with the exception of the waterless racing type. Water is best at absorbing more heat per gallon than any other liquid coolant, but if it was used alone it will cause corrosion in the engine. The purpose of antifreeze is to prevent the engine block from bursting due to expansion when water freezes.

Antifreeze is an additive that lowers the freezing point of a water-based liquid and also increases its boiling point. An antifreeze mixture is used to achieve freezing-point depression for cold weather and also achieves boiling-point elevation to allow higher engine temperatures. Freezing and boiling points are properties of a solution that depend on the concentration of the dissolved substance.

Commercially, both the additive (pure concentrate) and the mixture (diluted solution) are called antifreeze, depending on the context. Corrosion inhibitors are added to help protect the radiators and cylinder water jackets from corrosion. Water pump seal lubricant is also added to the coolant mix.

Careful selection of an antifreeze can enable a wide temperature range in which the mixture remains in the liquid phase, which is critical to efficient heat transfer and the proper functioning of heat exchangers.

Water

Water is the principal ingredient in coolant because it is inexpensive, is a very efficient heat exchange fluid, and has excellent thermal conductivity. Water also has a good specific heat. However, it freezes at 32°F (0°C) and boils at 212°F (100°C).

Water is the best coolant in terms of heat conduction, but it is also

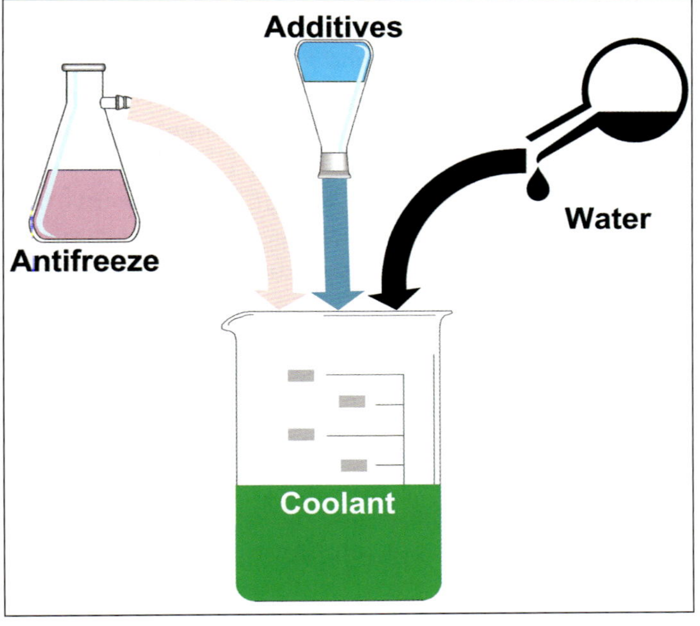

What is engine coolant? Coolant is a mixture of about 50 percent water, 47 percent antifreeze (ethylene glycol or other antifreeze), and 3 percent corrosion inhibitors and other additives.

Water Quality

The American Society for Testing and Materials (ASTM) standards for water quality are:

Chloride	Less than 40 ppm		Total Hardness	Less than 170 ppm
Sulfate	Less than 100 ppm		pH Range	5.5 to 9.0
Calcium	Less than 100 ppm		Iron	Less than 1 ppm
Magnesium	Less than 100 ppm			

* ppm = parts per million ■

the best source of corrosion. If you are using straight water, you should always add a water pump lubricant and a corrosion inhibitor. You can also use a coolant enhancer such as WaterWetter, which improves surface tension and heat conductivity.

Coolant manufacturers often suggest a 50-50 mix of ethylene glycol and water, which will protect your cooling system to −34°F. If using an ethylene glycol and water mix, it is also recommended to use distilled water to keep minerals out of the cooling system. Today, antifreeze already mixed with water can be purchased.

Ethylene Glycol

French chemist Charles-Adolphe Wurtz first synthesized ethylene glycol in 1856. However, not much was done with it until the 20th century. In 1917, ethylene glycol was first produced commercially for use in the manufacture of dynamite.

In addition to being an effective substitute for glycerol in the manufacture of high explosives, ethylene glycol also turned out to have characteristics that made it an ideal antifreeze coolant. It would mix readily with water in any proportion, and it has a lower freezing point and a higher boiling point than water. Ethylene glycol was first used as an automotive antifreeze in 1926.

Ethylene glycol solutions have been used as a permanent antifreeze

Ethylene glycol was the base antifreeze used in every OEM factory fill in the past. It was the familiar green coolant used in all makes and models of cars and trucks. The additive package varied by manufacturer. (Photo Courtesy Jim Halderman)

since the higher boiling points provided advantages for summertime use as well as during cold weather. Ethylene glycol oxidizes to five organic acids (formic, oxalic, glycolic, glyoxylic acid, and acetic acid). Inhibited ethylene glycol antifreeze mixes are available with additives that buffer the pH, slow corrosion to protect the engine metal, and reserve alkalinity of the solution to prevent oxidation of ethylene glycol and formation of

these acids. Nitrites, silicates, borates, and azoles may also be used to prevent corrosive attack on metal. Only the ethylene glycol is permanent; the corrosion inhibitor and other additives will wear out.

Ethylene glycol is poisonous to humans and other animals, and it should be handled carefully and disposed of properly. Its sweet taste can lead to accidental ingestion or allow its deliberate use as a murder weapon. Ethylene glycol is difficult to detect in the body. Its metabolism produces calcium oxalate, which crystallizes in the brain, heart, lungs, and kidneys, damaging them. Symptoms include intoxication, severe diarrhea, and vomiting. Depending on the level of exposure, accumulation of the poison in the body can last weeks or months before causing death, but death by acute kidney failure can result within 72 hours if the individual does not receive appropriate medical treatment for the poisoning.

Ethylene glycol was used for a variety of automotive applications, but there are lower-toxicity alternatives made with propylene glycol along with the newer organic hybrid antifreeze.

Embittered Ethylene Glycol

Newer ethylene glycol antifreeze mixtures contain the embittering agent denatonium benzoate to discourage accidental or deliberate con-

Embittered equals awful tasting. The embittering agent (denatonium benzoate, 30 ppm, has been required in California and Oregon since 2004. It prevents the death of animals that might drink ethylene glycol due to its sweet taste. (Photo Courtesy Jim Halderman)

sumption. Denatonium benzoate is very bitter. Just a small amount will make coolant so bitter that pets will not be able to swallow it.

There are 12 US states that require antifreeze to have the embittering agent. They are: Arizona, California, Maine, New Jersey, New Mexico, Oregon, Tennessee, Utah, Virginia, Vermont, Washington, and Wisconsin. Illinois, Massachusetts, New Hampshire, and Ohio are working on passing a similar rule.

Propylene Glycol

Propylene glycol is considerably less toxic than ethylene glycol and may be labeled as nontoxic antifreeze. It is used as antifreeze in situations where ethylene glycol would be inappropriate, such as in water pipes in homes.

Propylene glycol oxidizes when exposed to air and heat to form lactic acid. If not properly inhibited, it can be very corrosive, so pH buffering agents such as dipotassium phosphate and potassium bicarbonate are often added to prevent acidic corrosion of metal components. In the absence of inhibitors, propylene glycol can react with oxygen and metal ions, generating various compounds including organic acids (e.g., formic, oxalic, acetic). These acids corrode metals in the system.

To prevent corrosion, pre-inhibited propylene glycol solutions can also be used instead of pure propylene glycol. When a solution of propylene glycol in a cooling or heating system develops a reddish or

Propylene Glycol is sold by only a few companies, and one of the best versions is AMSOIL Propylene Glycol Antifreeze & Coolant (ANT). AMSOIL formulated ANT to provide benefits beyond those in conventional antifreeze and coolant products. Unlike toxic ethylene glycol–based products, AMSOIL Propylene Glycol Coolant has low toxicity, limiting the threat to children and animals. It is also biodegradable and provides maximum cooling system protection in extreme temperatures and operating conditions. The FDA has classified propylene glycol as an additive that is generally recognized as safe for use in food. (Photo Courtesy AMSOIL)

FDA Approval

The Federal Drug Administration (FDA) allows propylene glycol to be added to a large number of processed foods. It is found in ice cream, frozen custard, salad dressings, and baked goods along with the e-liquid used in electronic vapor cigarettes.

black color, this indicates that iron in the system is corroding and it should be replaced.

Some manufacturers do not recommend the use of propylene glycol antifreeze. Check the recommendation in the vehicle's owner's manual or online service information before using it.

Inorganic Additive Technology

Inorganic additive technology (IAT) is the traditional green coolant used in older vehicles. This solution can contain silicates (possible

Inorganic acid technology (IAT) contains either ethylene glycol or propylene glycol. It is mainly built up with silicate or phosphate additives to increase its compatibility with metal cooling system components. Most will recognize it as the original green or blue antifreeze. The recommended replacement interval is every year. (Photo Courtesy Jim Halderman)

abrasive dropouts), phosphates, and borates. It offers fast-acting corrosion protection, but the additives are quickly consumed. Once gone, the cooling system is exposed to possible corrosion problems, so it had to be changed regularly.

IAT is considered obsolete. It is no longer used because it can cause early failure of ceramic phenolic seals used in the newer water pumps.

Antifreeze Types

As the need for antifreeze grew, manufacturers became increasingly aware of the different needs for each region. They began to use different compounds and formulas to suit specific areas of the world.

European countries had extremely hard water. Since coolant is a 50-50 mix of antifreeze and water, water quality drastically impacts the overall mix. European manufacturers began to move away from phosphate-based technology because of its tendency to form scale.

On the other side of the world, Japanese manufacturers began to move away from silicates. They were avoiding issues with silicate gel drop out.

Organic Acid Technology

Organic acid technology (OAT) is an ethylene glycol–based antifreeze found in Dex-Cool. It is the coolant of choice for General Motors, Volkswagen, and many Japanese and Asian vehicles. This formula is engineered to offer long-life corrosion protection. It was originally made by Texaco and was first used in the GM 6.5L diesel engine. It was used exclusively by General Motors in some buildout 1995 models and all 1996 cars and trucks except Saturn and Geo.

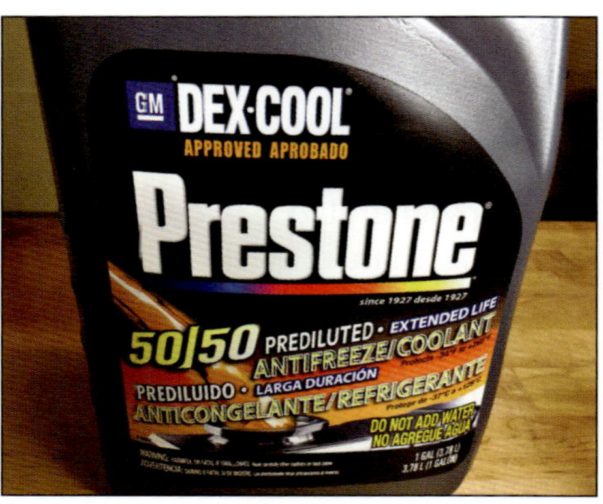

GM Dex-Cool's formula was engineered for corrosion protection. The downside of OAT is it is not compatible with other types of coolant (IAT and HOAT). In fact, Ford, Chrysler, and others say not to use this type of coolant in their newer vehicles.

Dex-Cool antifreeze uses two inhibitors: sebacate and ethylhexanoic acid (2-EH) as corrosive inhibitors. Dex-Cool has a bright orange color and its inhibitor formula forms a film on aluminum surfaces that protects these surfaces from corrosion. The disadvantage of using OAT is that it is not compatible with other types of coolant (IAT and HOAT). Also, 2-EH tends to damage plastics, such as nylon 6.6 that is used in intake manifold gaskets and radiators. There are two OAT coolants that do not use 2-EH: Peak Global OAT and G30 OAT.

There are a number of class action suits against General Motors, and they may have to pay for intake gaskets on V-6s and V-8 engines. Some of these vehicles exhibit brown gunk and rust in their cooling systems along with intake manifold gasket failures. The root cause of this brown gunk is air entering the system and causing rust. Rust in the system then causes blockages. Ford, Chrysler, and others say to not use Dex-Cool in their newer models.

Dex-Cool's bright orange color is very distinctive. It is not the green coolant that you grew up with! (Photo Courtesy Jim Halderman)

Prestone and Valvoline market organic acid technology (OAT) antifreeze coolants that meet the GM Dex-Cool specification, but they are not as expensive as the GM Dex-Cool-labeled products due to not having to pay royaltys to General Motors. Vehicles are built with OAT antifreeze (Dex-Cool) or with hybrid organic acid technology (HOAT) formulation (Zerex G-05), both of which claim to have an extended service life of 5 years or 150,000 miles.

Originally, it had a service interval of 150,000 miles or 5 years, but that has been downgraded to 2 years or 30,000 miles.

The real culprit appears to be operating vehicles for long periods

G-Coolants

Glysantin is an engine coolant created by BASF in Europe and Valvoline Zerex in the United States. These are the coolant types listed by G number:

G05: Different from Dex-Cool in certain amounts of additives

G30 and G34: Non-silicate and phosphate free

G11: Blue, Volkswagen before 1997

G12: Pink/red, Volkswagen 1997+ or purple, Volkswagen 2003+; HOAT formulation and phosphate free

G48: Blue, low silicate and nitrate, amines, and phosphate (NAP) free ■

of time with low coolant levels. The low coolant is caused by pressure caps that fail in the open position. This exposes hot engine components to air and vapors, causing corrosion and contamination of the coolant with iron oxide particles. This in turn can aggravate the pressure cap problem, as contamination holds the caps open permanently. New caps and recovery bottles were introduced at the same time as Dex-Cool. To avoid this issue, check the radiator pressure cap as outlined in chapter 7 and replace the cap if it won't hold the right pressure.

Prestone Extended Life is an OAT coolant that is compatible with any antifreeze that meets the Dex-Cool specification. It is less expensive than Dex-Cool products. According to some Dex-Cool manufacturers, mixing a green non-CAT coolant with Dex-Cool reduces the batch's change interval to 2 years or 30,000 miles but will not cause engine damage.

Honda and Toyota created new extended life coolants that use OAT with Sebacate but without the 2-EH. Some added phosphates provide protection while the OAT builds up. Honda specifically excludes 2-EH from its formulas. Typically, OAT antifreeze contains an orange dye to differentiate it from the conventional glycol-based coolants (green or yellow). Some of the newer OAT coolants claim to be compatible with all types of OAT and glycol-based coolants; these are typically green or yellow in color.

Hybrid Organic Acid Technology

A newer version of OAT technology is called hybrid organic acid technology (HOAT). It is similar to the OAT-type antifreeze in that it uses organic acid salts (carboxylates) that are not abrasive to water pumps yet provide the correct pH. If the pH is too high, the coolant can cause scaling and reduce the heat transferability of the coolant. If the pH is too low, the resulting acidic solution could cause corrosion of the engine components exposed to the coolant.

HOAT is a combination of IAT and OAT with nitrites that actually existed prior to the development of OAT. The generally recommended replacement interval is 3 years or 150,000 miles. (Photo Courtesy Jim Halderman)

HOAT is used in newer Ford, Chrysler, and Mercedes vehicles. It uses the best aspects of both IAT and OAT, making HOAT a very protective, long-life coolant. HOAT coolants typically mix an OAT with a traditional inhibitor, such as silicate or phosphate. HOAT can be dyed red, pink, yellow, or blue.

HOAT types are as follows:

Volkswagen or Audi: Pink; contains some silicates and an organic acid; phosphate free

Mercedes: Yellow; contains low amounts of silicate; phosphate free

Ford: Yellow; contains low silicate; phosphate free; dyed yellow for identification

Ford yellow HOAT (shown) contains low silicate, no phosphate, and is dyed yellow for identification.

Honda: Blue; contains a special coolant with just one organic acid

European or Korean: Blue; contains low silicates; phosphate free

Asian: Red; contains no silicates; has some phosphate

Phosphate Hybrid Organic Acid Technology

Phosphate hybrid organic acid technology (PHOAT) was developed for Mazda-based Fords of 2008 and later. It is ethylene glycol based and available in a 55-percent coolant and 45-percent water premix. The premix containers make sure that the water used meets specifications. The use of PHOAT coolant in Ford Mazda engines is required to be assured of proper protection of the internal coolant passages in the engine.

Phosphate hybrid organic acid technology (PHOAT) was developed for Mazda-based Fords in 2008 and later, such as the Mazda FL-22. It is ethylene glycol based. Ford Mazda engines required the use of PHOAT coolant to be assured of proper protection of the internal coolant passages. It is also only available in premix containers to make sure that the water used meets specifications. (Photo Courtesy Jim Halderman)

PHOAT Properties

Concentration: 55 percent

Boiling point (with 15-psi pressure cap): 270°F (132°C)

Freezing point: −47°F (−44°C)

Color: Dark green

Embittered (made to taste bitter so animals will not drink it)

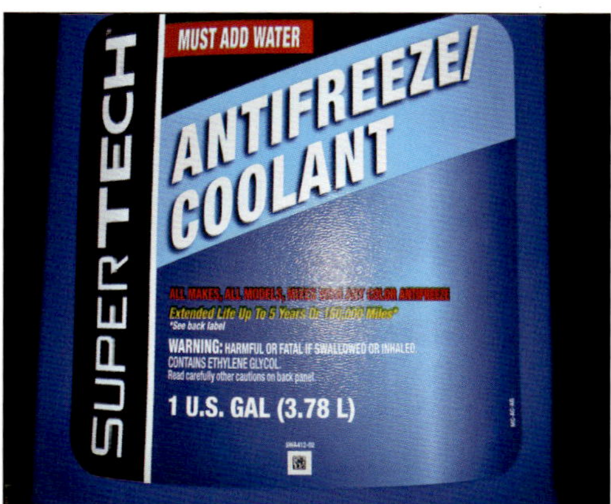

North American green antifreeze was the original "universal" formula, and it was the coolant used until the introduction of extended-life coolants. The fast-acting silicate and phosphate corrosion inhibitors provided quick protection for bare iron and aluminum surfaces and had a proven track record of providing trouble-free service in virtually any vehicle application (domestic, Asian, or European), assuming the chemistry is correct. OAT coolants should not be used in a vehicle that specifies the use of a hybrid OAT coolant. Always check the vehicle's owner's manual. Yet, the short-lived nature of the corrosion inhibitors means this type of coolant should be changed every 2 to 3 years or 30,000 miles.

Universal Coolant

Universal coolant is usually a HOAT antifreeze. It is extended life, low-silicate, phosphate-free coolant. It can be used in many vehicles. However, it cannot be use in Mazda vehicles and cannot meet the needs for engines requiring a silicate-free formula.

Antifreeze Components

Coolant is made up of an antifreeze component and a series of additives. Irrespective of the type of coolant and its color, the only difference between all the original equipment coolants is in the additives. This means about 97 percent of all coolants are the same; the only difference is in the additive package and color used to help identify the coolant.

Traditional Inhibitors

There were two major corrosion inhibitors traditionally used in vehicles: silicates and phosphates. American-made vehicles used both silicates and phosphates. European makes contain silicates and other inhibitors but no phosphates. Japanese vehicles traditionally use phosphates and other inhibitors, but no silicates.

Additives

All automotive antifreeze formulations, including the newer OAT and HOAT antifreezes, are environmentally hazardous because of the blend of additives, including lubricants, buffers, and corrosion inhibitors. Most commercial antifreeze formulations include corrosion-inhibiting compounds and a colored dye (commonly a fluorescent green, red, orange, yellow, or blue) to aid in identification.

Additives make up about 5 percent of the coolant. These additives are proprietary, so OEM safety data sheets (SDS) list only those compounds that are considered to be significant safety hazards when used in accordance with the manufacturer's recommendations. Common additives include sodium silicate, disodium phosphate, sodium molybdate, sodium borate, denatonium benzoate, and dextin (hydroxyethyl starch).

Disodium fluorescein dyes are added to antifreeze to help trace the source of leaks, and it is as an identifier because some formulations are incompatible. Automotive antifreeze has a characteristic odor due to the addition of the corrosion inhibitor tolytriazole.

Waterless Glycerol-Based Coolant

There are glycerol-based coolants that do not require water. Glycerol is nontoxic, noncorrosive, and withstands relatively high temperatures. It was used as an antifreeze for automotive applications due to its low freezing point before it was replaced by ethylene glycol.

Like ethylene glycol and propylene glycol, glycerol is a non-ionic kosmotrope that forms strong hydrogen bonds with water molecules, competing with water for the hydrogen bonds. This interrupts the formation of ice unless the temperature is significantly low. Glycerol is mandated for use as an antifreeze in many sprinkler systems.

There are very few waterless coolants that use glycerol.

Evans Coolant

Water provides very superior heat transfer, but only if it remains in its liquid state. Evans waterless coolant does its job of removing heat well past the failure point of 50-50 coolant. Vapor pressure is also reduced with Evans compared to 50-50 coolant.

Evans makes several different waterless coolants: High Performance, Powersports, Heavy Duty, and Non-Propylene Glycol (NPG). Its products can be used in high-performance street vehicles and work best with high-flow cooling systems. In most cases, the stock cooling system configuration is sufficient, and there is no need to change

Evans Heavy Duty waterless coolant is the original lifetime waterless coolant that does not require replacement. It is recommended for racing engines run on tracks or in series where there is a ban on ethylene glycol. (Photo Courtesy Evans Coolant)

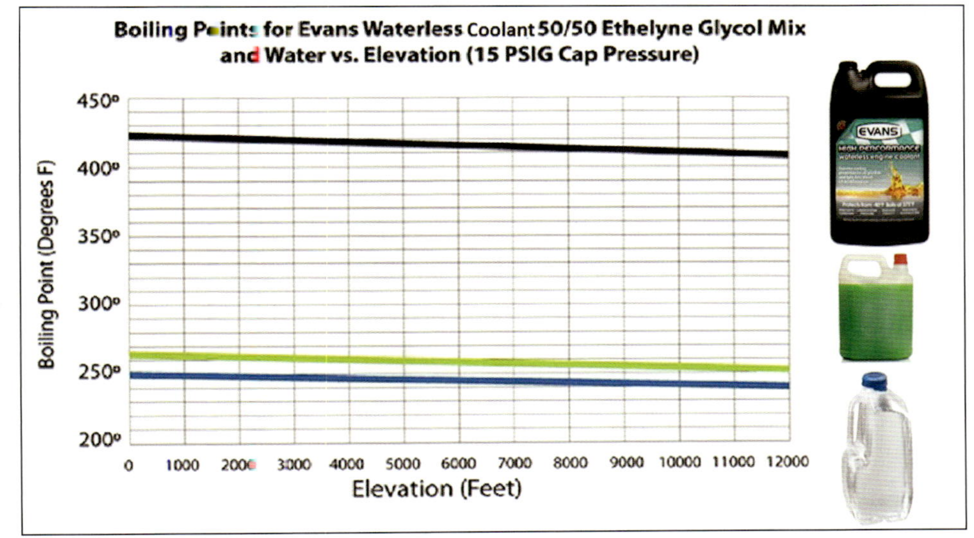

At 375°F, Evans' boiling point is more than 100 degrees higher than water-based coolant. Ethylene glycol is around 275°F and water's boiling point is 212°F. (Chart Courtesy Evans)

Questions and Answers with Evans Engineer Mike Tourville

Q What are the advantages of Evans waterless coolants?

A Evans contains no water, has a boiling point of 375°F, and freezes at –40°F. This means no vapor formation, no vapor pressure, and no boil over. Evans maintains its liquid state, and with constant liquid-to-metal contact, efficient heat transfer continues even under higher temperatures. Evans's lower operating pressure reduces stress on hoses, seals, and gaskets. The absence of water also means no corrosion and electrolysis.

Q How does Evans waterless coolant function differently in the cooling system?

A Evans avoids the boiling condition at stressed temperatures, which is the weakness of water. Evans waterless coolants remain liquid, maintain good heat transfer, keep engine component temperatures safe, reduce pressure, and avoid rust and galvanic erosion.

Q How do Evans waterless coolants control engine metal temperatures as compared to water-based coolants?

A Water-based coolant boils at a temperature only slightly higher than the operating temperature of the coolant. Localized boiling releases water vapor that can only condense into coolant that is colder than the boiling point of water. Vapor that doesn't condense occupies a volume that displaces liquid coolant. Hot engine metal, insulated by water vapor, becomes an engine hot spot that can cause pre-ignition and detonation. Evans's high boiling point means it will not turn to vapor.

Q Will Evans coolant lower the operating temperature of my engine?

A Vehicles running under normal operating conditions should show either no change or a slight increase in temperature, but that will depend on the cooling system configuration and driving conditions. Certain systems that use incompatible components, have an existing problem, or are poorly designed could run hotter. The ability to lower the operating temperature depends on multiple factors—primarily coolant flow volume and airflow temperature. For example, multi-pass radiators will result in higher temperatures due to decreased coolant flow volume versus large-tube multirow radiators that improve coolant flow. Different thermostats may increase flow volume because of less restriction.

Q Why does my engine run hotter after switching to Evans waterless coolants?

A A system that is highly optimized for water (with restrictive flow and high-pressure differentials) can cause slower circulation with Evans waterless coolants. Evans won't boil, but if it is held longer in the engine, it can pick up more heat. The coolant temperature is what your gauge reads. If the rest of the system is capable and compatible, Evans waterless coolants can process more heat out of the engine, and the engine component temps are actually improved and stabilized.

Q Why do museums and collectors such as Jay Leno use Evans?

A They use Evans for corrosion protection in older vehicles that have dissimilar metals in the system that sit for long periods of time. And of course, they use it for temperature control under the hot Southern California climate and traffic conditions.

Q How does Evans prevent water pump cavitation?

A Evans waterless coolant inhibits vapor development in the pump over a broad range of temperatures. With Evans waterless coolant, the suction side of the coolant pump is never at a low enough pressure to flash vaporize the coolant. The pump never gets vapor-bound and can continue to pump coolant. No vapor bubbles are formed to collapse against the metal and cause cavitation erosion damage to the pump.

Q How do I install Evans waterless coolant?

A The conversion process is not complicated, but it should be done thoroughly and according to written instructions, along with instructional videos on our website: evanscooling.com.

1. Drain all old water-based coolant from radiator, block, and heater core if accessible.
2. Use high-volume air to force out remaining coolant.
3. Fill with Evans Prep Fluid (waterless flush) and run for 15 minutes to circulate.
4. Allow to cool and drain the Prep Fluid in the same manner as the old water-based coolant.
5. Fill with Evans waterless coolant and run for 15 minutes to circulate.
6. Test for water content to confirm there is less than 5 percent water. Water content can be measured with a refractometer or a sample can be sent to Evans for testing.

Q How do I get all the water out? What if some water is left?

A Excess residual water can bring back the symptoms that Evans is intended to eliminate: the effects of vapor and corrosion. For best results, maintain a water percentage below 5 percent. If a loss of coolant occurs from a leak, requiring an emergency top off, water can be used and will mix with the Evans. The additional water will compromise the benefits of Evans, and it is recommended to reinstall Evans within a few weeks.

Q How much Evans Prep Fluid do I need to use?

A If you cannot fully drain the system, open the lower radiator hose, block drain (if accessible), and heater core. Allow them to empty and force high-volume air to purge the remaining coolant. Fill with Evans Prep Fluid and run the vehicle to circulate and drain again. This would require approximately 75 percent of the system volume of Prep. Alternately, smaller quantities of Prep can be used to flush through a component or plumbing.

Q What happens if I have water in my cooling system after installing Evans?

A A water content higher than 5 percent will lower the boiling point and may reduce the corrosion protection. If a water content exceeds 5 percent, drain a portion from the system and add back new Evans waterless coolant until it is below 5 percent.

Q If I have a leak and Evans is not immediately available what can I safely add?

A The likelihood of coolant loss and the need for topping up are greatly reduced with Evans. In the event a leak occurs and Evans is not available, water or water-based coolant may be used. The cooling system will function but the boil-over and corrosion protection of Evans waterless coolant will be reduced. As a temporary measure, stop leak products may be used. Current approved dry-type stop leaks are Bars Leak tablet part #HDC, GM Cooling system Tabs, Aluma Seal, and Copper Seal. Liquid stop leak products compatible with Evans are: Bars Leak part number 1186 and K-Seal part number ST5501.

Q Do I need to change my radiator cap when using Evans?

A No, a different radiator pressure cap is not required. Evans waterless coolant expands slightly as it warms, creating pressure of 3 to 5 psi, and the existing cap does not need to be changed.

Q What other system modifications are recommended to optimize Evans waterless coolant?

A In most cases, no modifications are needed; however, the higher coolant flow rate the better. This can include increasing pump output and lessening restrictions such as the thermostat. No need to change the pressure cap, as the point at which the cap will vent is less important, as Evans waterless coolant does not require pressure to keep it in the system. Contact Evans's tech support team for recommendations for your system at 888-990-2665.

Q Can I use it with a multi-pass radiator?

A Multi-pass radiators are relatively restrictive. These radiators work best in water-based systems, which use pressure-differential to assist circulation. Evans waterless coolant, without the pressure, works best in high-flow, single-pass core designs.

Q What type of radiator does Evans recommend?

A Evans recommends single-pass radiators as they have less flow resistance than multi-pass radiators. The following are minimum radiator core suggestions:

300 hp or less without AC ...4 rows: 0.5-inch tube copper/brass
300 hp to 400 hp with AC ...2 rows: 1-inch tube aluminum
400 hp to 600 hp.................2 rows: 1.25-inch tube, aluminum
600 hp and above...............3 rows: 1-inch tube aluminum
600 hp and above...............2 rows: 1.5-inch tube aluminum

Q Does Evans require periodic maintenance?

A No periodic addition of additives is required, nor should any ever be added. Evans recommends inspecting the cooling system at least once a year to ensure water content remains below 5 percent. ■

the thermostat or radiator pressure cap. Operation system modifications, such as increasing coolant flow rate, can be done to optimize Evans's performance.

At temperatures toward −40°F, Evans may become thicker. This means it will contract, as opposed to water, which expands when it freezes. Because it contracts, Evans prevents freeze burst damage to cooling system plumbing. That is why it is not suggested for street-driven vehicles.

Evans waterless coolant has a high boiling point, which means it can withstand relatively high temperatures without fear of forming vapor, building pressure, or boiling out. It also prevents electrolysis and corrosion. Evans waterless coolant boiling point is 375°F, which is more than 100 degrees higher than water-based coolant.

Evans High Performance waterless coolant is used straight with no water added. It can last a lifetime; it does not require replacement as long as it does not become contaminated with excess water (less than 5-percent water is optimal). Its anticorrosive properties and high boiling point make it ideal for muscle cars, classic cars, hot rods, and motorcycles. Evans is also used in racing venues, particularly in powersports. It is a contingency sponsor for the National Hot Rod Association (NHRA) and is authorized for use at NHRA tracks.

Before introducing Evans, it is recommended to make sure all hoses, seals, and gaskets are leak-free. It is also important to remove all of the old 50-50 coolant (water/ethylene glycol) to ensure water content is below 3 percent. Evans offers a Prep Fluid as a waterless flush to assist with the removal of the old coolant.

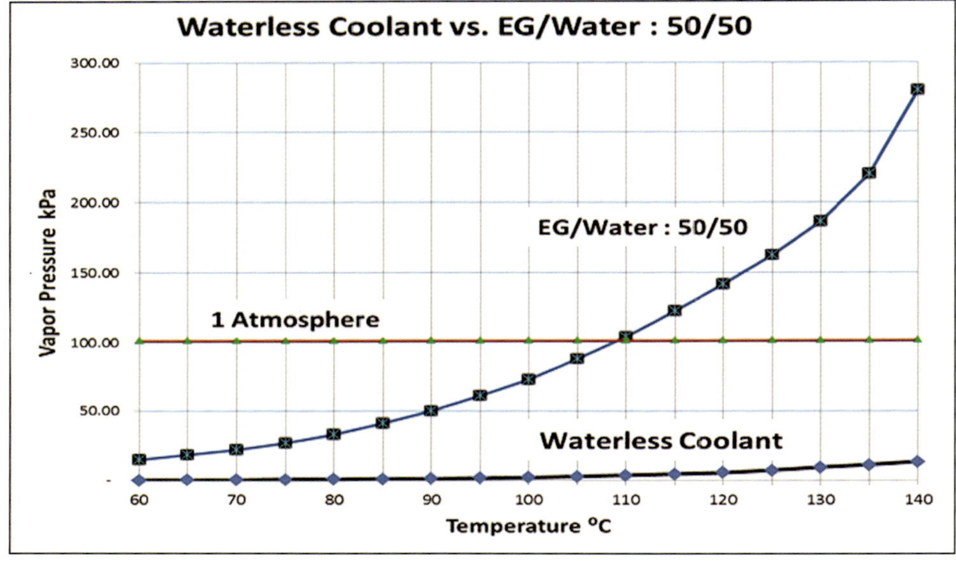

The red line in this graph shows thermal conductivity compared to temperature in degrees Celsius of a coolant mixture of ethylene glycol and water. The blue line shows thermal conductivity compared to temperature in degrees Celsius of Evans waterless coolant. Water provides very superior heat transfer when it remains in a liquid state. Evans waterless coolant will remove heat far past the failure point of 50-percent water and 50-percent ethylene glycol mix. (Chart Courtesy Evans)

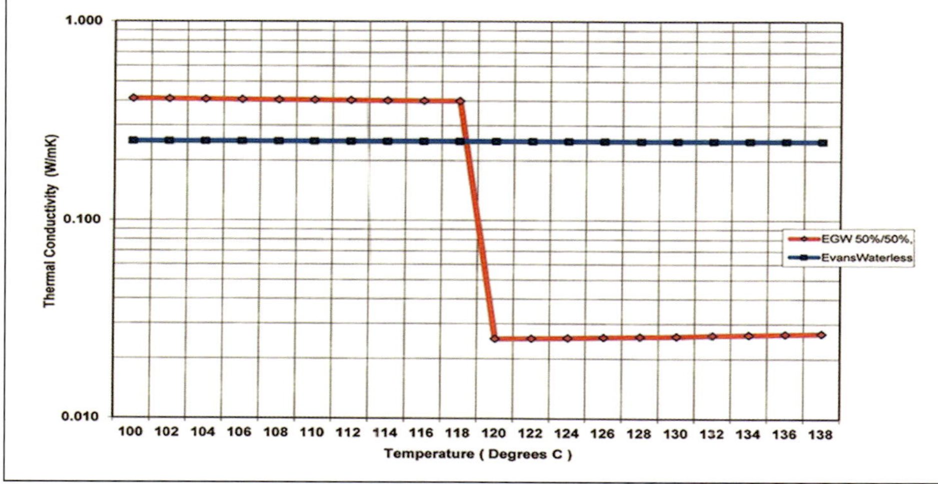

This graph compares Evans waterless coolant in the blue line with squares to a 50-50 mix of ethylene glycol and water-based coolant in black line with dots against a flat line pressure of one atmosphere to show that the vapor pressure is significantly less using the waterless coolant. (Chart Courtesy Evans)

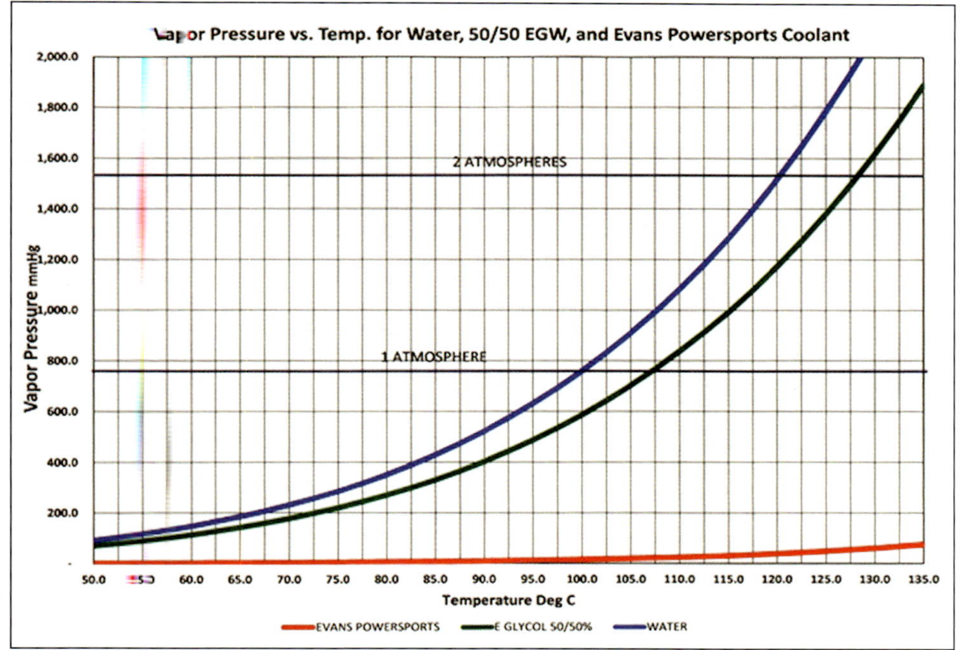

This graph demonstrates how the use of Evans waterless coolant alone reduces cooling system pressure as compared to 50-50 mix coolant. Temperature changes are shown in the red line. Flat line pressure at the 1 and 2 atmospheres are also shown on this graph. (Chart Courtesy Evans)

BeCool specific engine coolant uses propylene glycol.

BeCool Coolant

The BeCool Company offers Be Coolant, which is an extended life, super-duty coolant used for all stock, modified stock, and racing applications. Be Coolant features a new-generation earth-friendly biodegradable propylene glycol formula that protects cooling systems and aluminum components down to −26°F.

It provides 300,000 miles of extended-life coolant protection for all performance and severe-duty applications including American, European, and Asian high-performance cars, as well as diesel-powered tow vehicles and recreational vehicles. Be Coolant features a self-sealing capability that is ideal for sealing hairline cracks in aluminum blocks and heads.

Coolant Service

Coolant service is a typical do-it-yourself task because the processes are not difficult and the tools needed are not expensive. First perform a visual inspection to see if the coolant is clean and bright and the system is full and free of air. With just a few simple tools, check the freezing and boiling points. When you have a high freezing point or low boiling point, there is too much water in the coolant.

In the absence of leaks, antifreeze chemicals such as ethylene glycol or propylene glycol may retain their basic properties indefinitely. By contrast, corrosion inhibitors are gradually used up, and they must be replenished from time to time. Larger systems (such as HVAC systems) are often monitored by specialist firms that take responsibility for adding corrosion inhibitors and regulating coolant composition. For simplicity, most automotive manufacturers recommend periodic complete replacement of engine coolant to renew corrosion inhibitors and remove accumulated contaminants. See chapter 7 for complete information on the maintenance and testing of engine coolant.

RADIATORS

A radiator is a heat exchanger that lowers the temperature of the liquid coolant after it leaves the internal combustion engine. Radiators are made from header and collector tanks linked by a core with many narrow passages that provide a high surface area relative to volume. A radiator cools the coolant that has absorbed the heat energy from the engine. A radiator does not return the temperature of the coolant back to ambient air temperature, but it is enough of a transfer to keep the engine from overheating.

Up until the 1980s, radiator cores were primarily made from copper (for fins) and brass (for tubes, headers, tanks, and side plates). Starting in the 1970s, the use of aluminum increased due to its reduced weight and cost. Eventually, aluminum took over the vast majority of vehicular radiator applications. Modern radiator cores are usually made of stacked layers of metal sheeting that has been pressed to form channels. The channels are soldered or brazed together.

The header tank of the radiator is located either on the top of the radiator or along one side. Hot coolant is fed into the header tank, distributed through the radiator core through tubes, and passes to the collector tank on the opposite end of the radiator. As the coolant passes through the radiator tubes, it transfers much of its heat to the tubes, which transfers the heat to the cooling fins that are between each row of tubes. The cooling fins release the heat to the ambient air. Cooling fins are used to increase the contact surface of the tubes to the air, thus increasing heat transfer efficiency. The cooled coolant is fed back to the engine, and the cycle repeats.

Since air has a lower heat capacity and density than liquid coolants, a fairly large volume flow rate must be moved through the radiator core to absorb the heat from the coolant. Radiators often have one or more fans that blow or pull air through the radiator. To save fan power

This photo shows a high-performance down-flow design radiator. It uses dual cooling fans that have their own ring design shrouds to help direct airflow through the radiator. (Photo Courtesy Champion Cooling Systems)

This modified Corvette has an electric cooling fan located behind the radiator and in front of the water pump. The fan uses a flush fit shroud, which helps increase airflow though the radiator.

Did You Know?

The US interstate system was first built in the western states in the early 1950s and 1960s. This highway system was designed with emergency sidings so that drivers could pull over and cool down their engines. These emergency pullover areas were equipped with water barrels so drivers could refill the over-heated radiators.

Thermosiphon Cooling Systems

Early motor vehicles used thermosiphon circulation to move cooling water between the cylinder block and the radiator. This system depended on forward movement of the car and fans to move enough air through the radiator to provide the temperature differential that caused the thermosiphon circulation. As engine power increased, an increase in flow was required, so engine-driven pumps were added to assist circulation.

More compact engines used smaller radiators and required more convoluted flow patterns, so the circulation became entirely dependent on the pump. It might even be reversed against the natural circulation.

An engine cooled only by thermosiphon is susceptible to overheating during prolonged periods of idling or very slow travel. In these instances, airflow through the radiator is limited unless one or more fans are able to move enough air to provide adequate cooling. These systems are also very sensitive to low coolant levels because losing even a small amount of coolant stops the circulation.

Henry Ford built his first car in 1896 with the thermosiphon system. It used a flat tank that was located under the driver's seat, which kept the seat warm in the wintertime. This early engine cooling system was very inefficient, and overheating and coolant loss was very common. Later, Ford added a thin heat exchanger connected to a tank of coolant, which became known as the automotive radiator. ■

The graphic shows a representation of the thermosiphon cooling system circulation from the radiator to the engine block. It relied on the forward movement of the vehicle. The fan helped to move enough air through the radiator to cause difference in temperature between the radiator and the engine block, causing the thermosiphon effect.

This Ford Model T used thermosiphon circulation.

consumption, radiators are often behind the front grille. This position aids in cooling because a fan in front of the radiator would impede airflow. Ram air can give a portion or all of the necessary cooling airflow when the coolant temperature remains below the system's designed maximum temperature and the fan remains disengaged.

Radiator Design

A properly designed cooling system consists of the right size radiator, a water pump with a good flow rate,

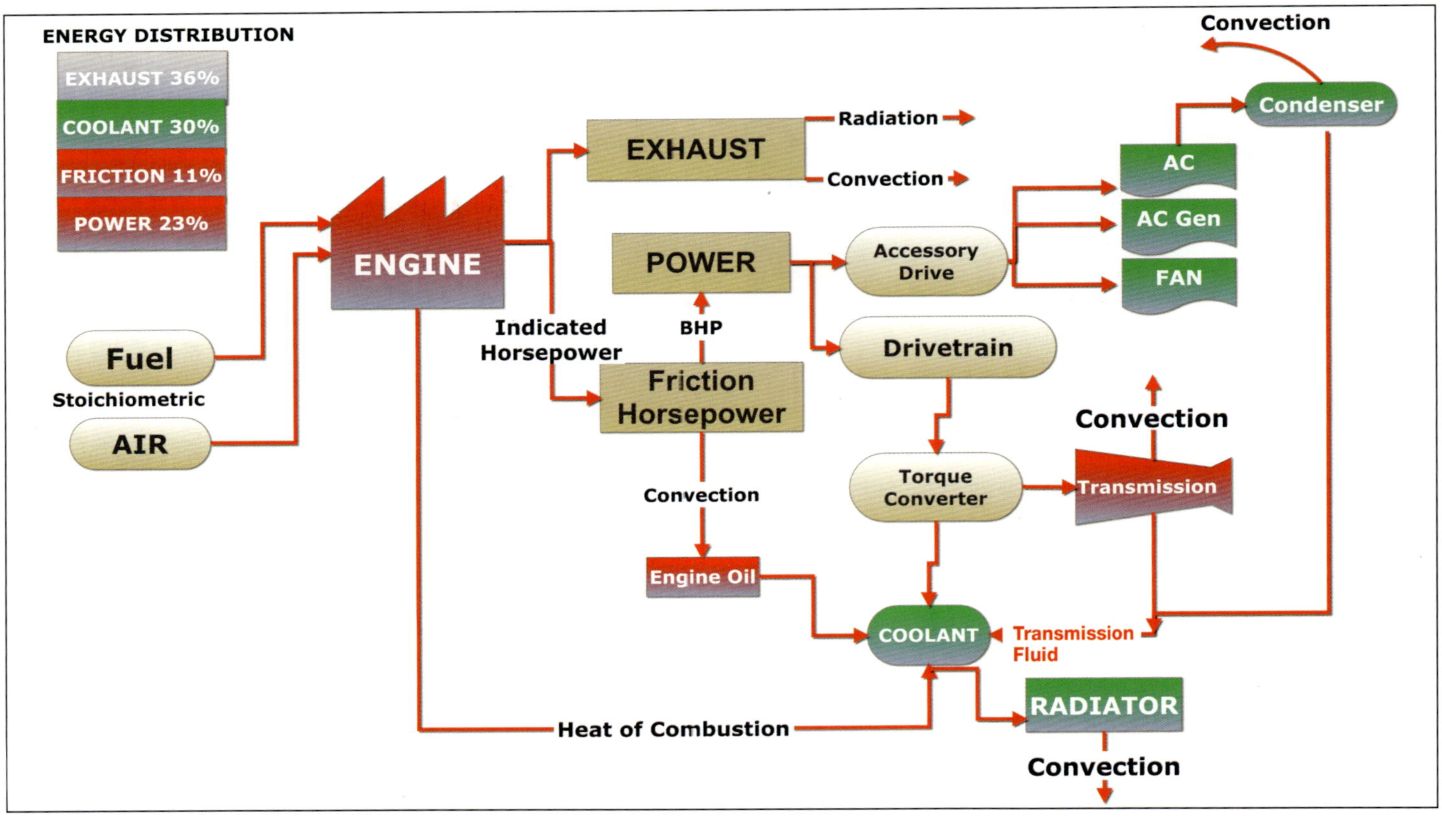

This flow diagram shows all of the items to consider when designing a radiator. They include engine horsepower, road speed, loads the vehicle will be placed under, and the airflow through the radiator, including the air-conditioner condenser.

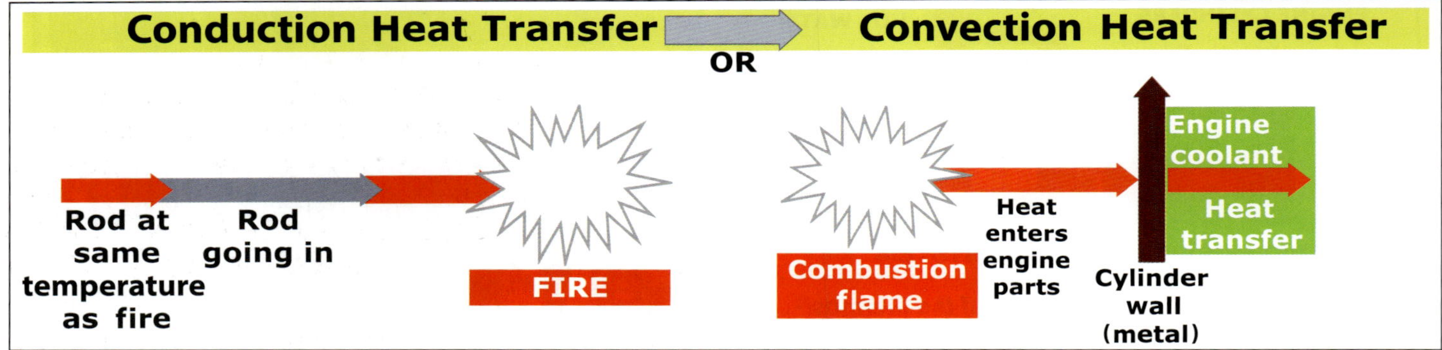

Conduction (left) is the transfer of heat from molecule to molecule through solids and fluids in intimate contact at rest. The iron rod is in the fire, and the end you are holding will become warmer and warmer because the heat travels through the rod via conduction from molecule to molecule, until the end you are holding is near the temperature of the end in the fire. Convection (right) is heat transfer by the molecular motion in the heated substance itself and only takes place in liquids and gases. The heat is transferred to the cylinder wall from heat of combustion where it goes into coolant and is carried away at the radiator.

and a cooling fan (either mechanical or electric) that provides enough airflow at idle or low RPM. At highway speeds, the air flowing through your radiator will keep your engine at operating temperature.

The radiator is located between the cooling fan and the vehicle header panel. The radiator removes heat from the coolant by conduction and convection, which is explained in detail in chapter 1. Conduction is the transfer of heat from molecule to molecule through solids and fluids that are in contact. Convection is heat transfer by the molecular motion in the heated substance itself and only takes place in liquids and gases. It is heat that is transferred to the cylinder wall from combustion, goes into coolant, and is carried away to the radiator.

Radiator Core

Heated coolant circulates in the radiator core, where coolant flows through coolant tubes. The tubes are exposed to the air flowing through the core, which has metal cooling fins between the tubes. Heat is transferred through the coolant tube walls along with the soldered joint to the cooling fins. These cooling fins are usually oval shaped to create turbulence in their airways, which helps transfer the heat between the core metal surfaces and the air flowing through the radiator. The air that flows through the radiator removes heat from the coolant and carries it away.

Coolant tubes are created from 0.0045 to 0.012 inch (0.1 to 0.3 mm) sheet brass or aluminum, using the thinnest possible materials for each application. This material is rolled into round tubes and the joints are sealed with a locking seam.

This illustration shows a radiator core design with horizontal cooling fins and vertical coolant tubes. Air passes through the cooling fins, rotating them in a turbulent fashion. This pulls heat out of the coolant that is flowing through the vertical coolant tubes.

Cellular or Film Design Core

Early radiators, such as those used in a Model T, were shaped like a honeycomb. They consisted of a stack of air tubes with bulging ends that were sealed together so liquid could continuously flow between these tubes. This radiator design allowed for some very fancy grilles and ornate radiators similar to those used on the Model T Ford.

This cooling system was inefficient and lacked a secondary heat transfer surface, so the honeycomb radiator design eventually went away. Some vintage cars use radiator cores made from coiled tube, a less efficient but simpler construction.

Early radiators used a honeycomb design. Round tubes were swaged into hexagons at their ends, then stacked together and soldered. As they only touched at their ends, this formed what became in effect a solid water tank with many air tubes through it.

Another early radiator core design was called the cellular or film design, which used zigzag ribbon airways made from copper as a secondary heat transfer surface. This type of matrix construction was once widely used. It generally provided a high rate of heat dissipation with a minimum radiator weight. Cellular or film design led the radiator design toward the pressurized cooling system. (Photo Courtesy Jim Halderman)

The honeycomb design was replaced by the cellular or film design. In this design, the airways and secondary heat transfer surfaces were formed by a series of zigzag ribbons made out of copper. Each airway row was attached with a corrugated copper strip. The inward facing peaks of the corrugations acted as spacers for the air fins, while the outward facing peaks were dented across part of their width to form liquid coolant channels. This design was used because it dissipated heat quickly in a very small, light radiator.

Tube and Flat Fin Design Core

The next design used was the tube and flat fin. It used a series of inline or staggered coolant tubes that were assembled through a stack of continuous air fins spaced closely together. These air fins acted as the secondary heat transfer surface.

The coolant tubes were made

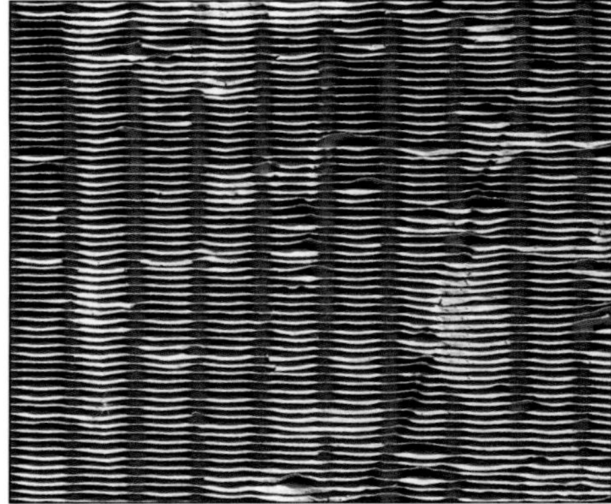

The tube and flat fin design radiator was another radiator core using inline vertical coolant tubes with horizontal cooling fins that ran right to left. This design had increased durability and eliminated solder corrosion. Later applications of the tube and fin design used the controlled atmosphere brazing (CAB) process, resulting in much stronger tube-to-header joints compared to copper-brass radiators.

from thin brass strips that had a flattened oval appearance. Thinner copper strips were used for the air fins. Louvers were used to promote turbulence in the airways. This core design was stronger than the early cellular or film design and was very significant during the 1940s, when the pressurized cooling system was introduced.

Corrugated Fin Design

The corrugated fin design (also known as the pack type) is the most common system. It is primarily used in today's radiators for street, modified street, and race cars. Corrugated fin radiators are manufactured by both original and aftermarket manufacturers. This design uses alternate rows of brass or aluminum coolant tubes and airways made of zigzag ribbons of copper or aluminum.

The coolant tubes are made of a flat section, and the cooling fins may be louvered to promote turbulence. There may be two or more rows of coolant tubes, but the demand for lighter vehicles has led to the development of single-row versions. These have a finer pitch with less separation for corrugated fins to maintain

a heat exchange capacity. This is because the heat radiated from the cooling fins is greater than that from the coolant tubes. This design is a conciliation between the cellular or film design and the tube and the flat fin design.

Radiator Materials

Originally, radiators cores were made from copper, and the collector tanks or header tanks were made of brass and bonded with solder. Copper and brass are very good materials for transferring heat, but they were heavier and are not renewable. This design was weak and the soldering usually worked loose from vibration and created leaks. So the copper and brass materials eventually gave way to aluminum.

Aluminum had been used for exhaust headers and collector tanks in commercial vehicles but not for radiators. In 1960, the Harrison Radiator Division of General Motors developed an aluminum radiator for the Chevrolet Corvette. This aluminum radiator used plastic tanks for production vehicles. The plastic tanks were lightweight, but they tended to

The corrugated fin design (also known as the pack type) is widely used in today's modern radiators. This design uses alternate rows of coolant tubes and airways made of aluminum tubes and zigzag ribbons of aluminum. (Photo Courtesy Champion Cooling Systems)

All radiators used in today's production vehicles are made out of aluminum, and many aftermarket ones are as well, such as this single pass Circle Track radiator from BeCool. Due to its light weight, it helps fuel economy and provides higher strength than copper or brass. (Photo Courtesy BeCool Performance)

crack, causing air (the enemy of cooling systems) to get in the system.

Today, auto manufacturers, including General Motors, Ford, Fiat-Chrysler and Toyota, use aluminum for radiators primarily to reduce weight. Aluminum is not as efficient as copper for heat transfer, but the reduced weight and strength allows for a larger, more-efficient radiator.

Radiator Types

Automotive radiators have two basic types: downflow and crossflow. Currently both types are used for stock, modified stock, and some racing applications. There are also dual-pass and tri-flow radiators.

Downflow Radiator

The downflow radiator was commonly used in older vehicles, and it is still used in modified stock. The radiator core is attached to an upper header tank and a lower collection tank. The header tank takes care of the expanding coolant and provides a reservoir in case any coolant is lost. The collector tank collects coolant that is returned to the engine after the heat has been removed and is then pumped back into the engine.

The downflow radiator design can use a remote header or overflow tank that is mounted separately from the radiator. This guarantees that the radiator will always be filled with coolant. The header tank also reduces any mixing of air with the coolant entering the radiator.

The overflow tank is a reservoir for engine coolant that is heated to its boiling point and would otherwise rise up and blow out of the radiator. The circulating coolant is separated from the air in the remote overflow tank. The overflow tank creates a closed cooling system that is more reliable and useful

Modified street vehicles often use a downflow radiator design. This design can use a header tank mounted separately from the radiator, which ensures that the radiator is always filled with coolant. It also reduces any mixing of air with the coolant entering the radiator because the aeration of the coolant has a bad effect on heat transfer efficiency. (Photo Courtesy Champion Cooling Systems)

than previous designs. As the temperature and pressure rise to the coolant's boiling point under pressure, the cap spring activates and allows the heated liquid to flow up and into the overflow tank. The system is under pressure so the liquid will reach a higher temperature before boiling, further increasing the effectiveness of the system.

This overflow tank can also reduce the possibility of water pump cavitation erosion caused by low pressure at the water pump inlet. A filler neck together with an overflow pipe connection are generally provided on the overflow tank. The overflow tank can be a standalone system on an older vehicle that does not use an expansion tank; on newer systems, it can be used in conjunction with the expansion tank.

Crossflow Radiator

The crossflow radiator was developed because the automotive industry needed a radiator design that had a core that was low and wide enough to provide airflow through the lower grilles. General Motors invented the crossflow radiator in the mid-1960s to compensate for these new grille designs. It was first used in the Chevrolet Corvette.

This radiator had an inlet (header) tank on one side and an outlet (collector) tank on the other side to provide the radiator with the reserve of coolant and prevent aeration. In some applications, a separate header tank was connected to the upper end of the outlet tank. In other applications, the separate header tank was eliminated. To prevent aeration, the filler cap and the overflow parts are located at the upper end of the outlet tank because that is where air tends to collect.

Dual-Pass Radiator

The dual-pass radiator has the coolant pass through the radiator core twice before returning it to the

A dual-pass radiator has a baffle welded inside the end tank. This baffle cuts the radiator in half so the coolant flows through each half in a series. Since each section is half the size of a full core, the coolant velocity is twice as fast and the pressure drop is now doubled. This design allows the coolant to stay in the radiator longer. The coolant has twice as far to go but it is also traveling twice as fast. The coolant flows through each tube only one time and would be like a U-flow in a manometer. The design has both the inlet and outlet on the same tank. (Photo Courtesy Champion Cooling Systems)

The crossflow radiator design has a collector tank on the passenger's side of the vehicle. It contains the automatic transmission fluid cooler along with the radiator pressure cap. The header tank is connected to the engine coolant outlet on the driver's side of the vehicle. (Photos Courtesy Champion Cooling Systems)

engine at a lower temperature. This design allows the engine coolant to go through two thermal transfer cycles, lowering the liquid in versus liquid out temperature to a much larger extent than a single-pass radiator.

The dual-pass radiator is identified by the placement of the inlet and outlet pipes on one side of the radiator. You can also see the weld marks between the separate units. The dual-pass radiator provides a lower output temperature.

Tri-Flow Radiator

Traditional radiators only pass the coolant one time in the front of the fan. Eastwood manufactures a triple-pass radiator called the tri-flow radiator. It has the coolant pass through the radiator core three times to provide lower outlet temperatures.

The tri-flow radiator has two rows and a 1-inch oval tube core. Its unique coolant path routes the coolant through the radiator core three times. The path increases turbulent flow while passing the coolant in

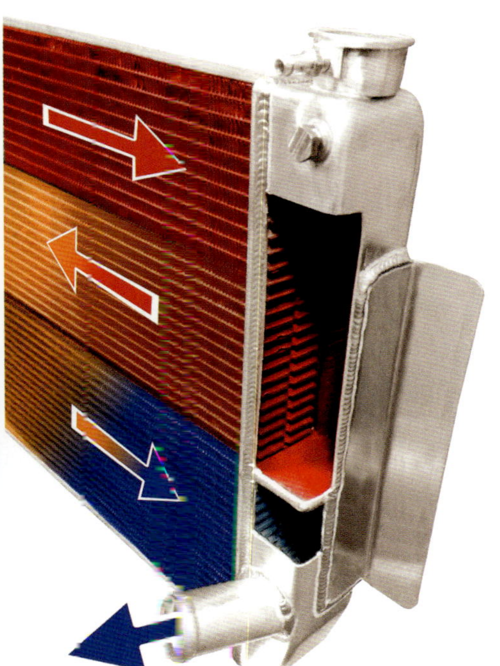

The triple-pass radiator, called the Tri-Flow radiator, has the coolant pass through the radiator core three times to provide lower outlet temperatures. The engine coolant goes through three thermal transfer cycles and lowers the liquid in versus liquid out temperature to a much larger extent than a dual-pass radiator design. (Printed by permission from Eastwood Company)

front of the fan three times. This system can help in some overheating situations.

Pressurized Cooling System

A limitation of most cooling systems is that the coolant should not be allowed to boil. The maximum amount of heat transfer is limited by the specific heat capacity of water and the difference in temperature between ambient temperature and the boiling point of water at 212°F (100°C). The thermal efficiency of the internal combustion engine increases with internal temperature, so the coolant must be kept at higher-than-atmospheric pressure to increase its boiling point.

When you pressurize the cooling system, the coolant can circulate at a higher boiling temperature, so that heat will be transferred more rapidly from the radiator through a greater or higher temperature differential between the two. This design will compensate for the lower atmospheric pressure (less than 14.7 psi) in high-altitude areas such as Denver, where the boiling point of the coolant would be lower and an overheat condition would be more likely.

The transfer of heat from the radiator is directly proportional to the temperature difference between the coolant and the radiator, so a pressurized system allows a radiator to be smaller but its design must be stronger to compensate. The increase in pressure at the pump inlet reduces the possibility of cavitation damage, as was previously discussed. Because the system has to be sealed, there is less coolant lost through evaporation and surging than if it were just vented.

Pressurized systems are more complex than vented systems. They are also more susceptible to damage because the coolant is under pressure. For example, minor damage in one of the radiator coolant tubes from a small puncture would cause the coolant to rapidly spray out of the hole. Failures of the cooling systems are one of the leading causes of engine failure.

Radiator Pressure Cap

John Karmazin of GM's Harrison Radiator Division invented the radiator pressure cap. GM's Buick Motor Division was the first to use the cap in 1939. The radiator pressure cap is a combination of a filler cap and a pressure control valve along with a vacuum control valve. It is placed at the highest point in the cooling system and seals against a seat in the filler neck of the radiator.

While the engine is running and the radiator pressure cap is in position, the cooling system will pressurize by pushing against a spring. This action takes place automatically when the coolant expands as it gets hotter. The reason for using this is to maintain the cooling system at a pressure that is above that of atmospheric pressure.

The radiator cap allows the engine's coolant to expand and contract without allowing air to enter

This radiator pressure cap is on a closed cooling system with a 15 psi opening pressure. It is on a late-model vehicle that specifies Dex-Cool OAT coolant and also has the instructions in Spanish. (Photo Courtesy Jim Halderman)

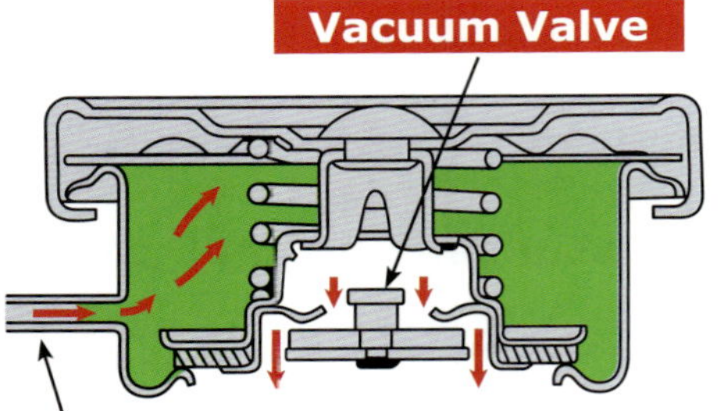

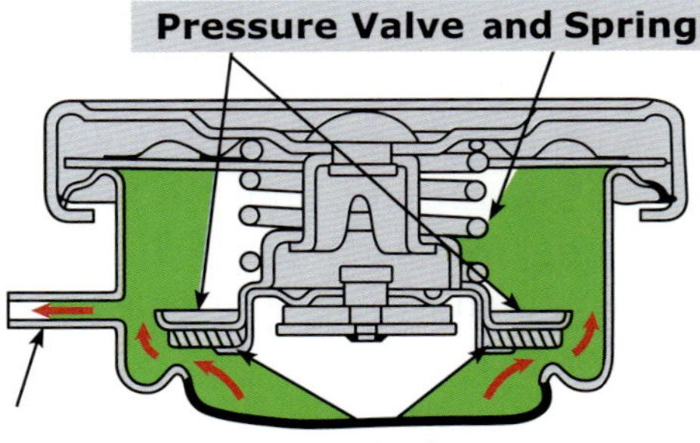

Radiator Cap with Vacuum Valve & Pressure Valve

Valve opens when pressure is low in system and allows coolant to return to system from recovery tank

Pressure increases in the cooling system by pushing against the spring pressure in cap. When pressure reaches cap pressure, valve opens and returns coolant to recovery tank.

The radiator pressure cap uses a pressure valve controlled by a spring that holds pressure in the system until it reaches a specified pressure and then opens. It also contains a vacuum valve. As temperatures drop and the coolant contracts, a vacuum opens and allows coolant to flow from the overflow tank back into the radiator. As pressure inside the system drops, outside air pressure helps the coolant flow in.

the cooling system. The upper seal protects the system at all times. After the engine warms and system pressure reaches the cap's rated pressure, the spring compresses and coolant flows into the reservoir or coolant overflow tank. This allows for expansion of the heated fluid.

Water always boils at 212°F under standard atmospheric pressure (14.7 psi). If the coolant is under pressure greater than 14.7 psi, the boiling point will be higher. It will increase as the pressure increases. This occurs because the coolant molecules are compressed by the pressure and will have to vibrate more for the temperature to increase.

The radiator pressure cap increases the cooling system pressure, which increases the boiling point and prevents coolant loss. For every pound of spring pressure, the boiling point is increased by 3 degrees. Water boils at 212°F, and increasing the pressure in a closed system increases the boiling point beyond that. A cap rated at 15 pounds will increase the boiling point in a system by 45°F with a boiling point of 257°F (212 + 45). High-performance radiator pressure caps range from 19 to 32 psi. Original manufacturers can design engines with higher operating temperatures.

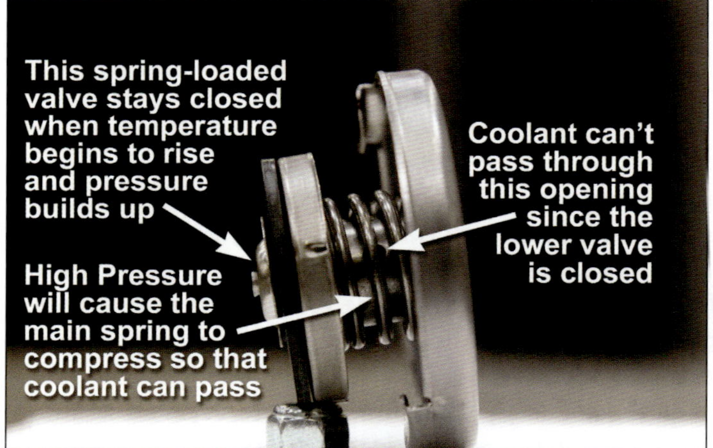

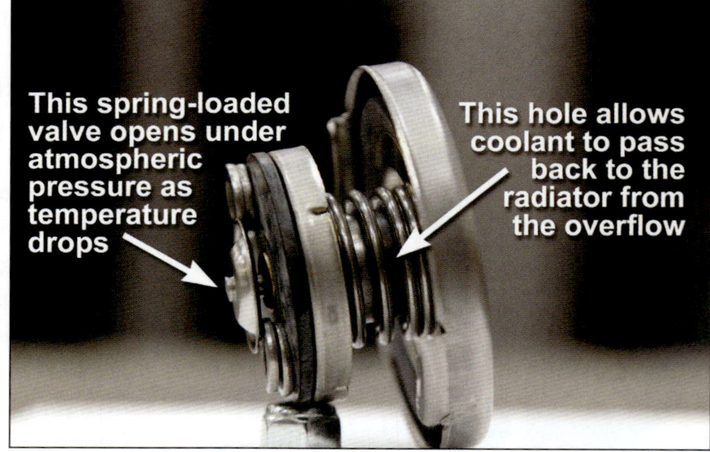

In the radiator pressure cap pressure valve operation (left), the spring-loaded valve stays closed until the opening temperature rises high enough to compress the spring and open the valve so coolant can pass through. The spring-loaded vacuum valve (right) opens when the pressure drops below 14.7 psi. (Photos Courtesy Champion Cooling Systems)

There are aftermarket BeCool high-performance radiator pressure caps rated at 21 to 25 psi (left) and at 28 to 32 psi (right). (Photos Courtesy BeCool Performance)

Radiator Cap Upgrades

Aftermarket radiators are upgrades to the OEM units, so you should pay attention to your radiator cap pressure rating. The coolant is under pressure to keep the boiling point as high as possible. That is why you want the highest pressure cap rating suitable for your application. Caps for older vehicles will most likely be rated for 7 to 12 pounds; newer vehicles will generally be in the 16 to 20 psi range.

If there is a leak in your system, higher pressure will push it out faster. The automotive cooling system is rated to a certain pressure by an extensive design process. The radiator pressure cap is designed to be the weak point in your cooling system so it can safely vent pressure. You don't want to use a cap that is so resistant to venting pressure that it causes some other part of the system to become the weak point. The higher the system pressure the greater the stress on the entire system, particularly gaskets, seals, junctions, and seams. If you go very high above the OEM's design pressure and the conditions are right, the hoses and even the radiator could burst.

If your cooling system is insufficient, it should be properly repaired and upgraded as necessary. Keep in mind that increasing the pressure beyond a couple of pounds above original spec is just not a good idea. The highest cap I have ever seen was 32 psi. In severe racing environments, you might see a higher pressure cap where you would have AN fittings and stainless steel–reinforced hoses, but I would follow the original designers' cap pressure, which is based on the original system design. Contact the manufacturer who makes the cap or radiator to see what the qualifying points are for its radiator pressure cap to determine the integrity of your cooling system.

Closed Cooling System

In a closed cooling system, it is not necessary to inspect the coolant level as often as an open system because it is sealed. This is different than the earlier nonsealed system, where the coolant would simply drain to the ground from the overflow pipe just below the pressure cap. The closed cooling system uses an expansion tank, which is part of the pressurized section of the cooling system. It allows about 1 gallon of extra cooling system capacity.

The expansion tank is under pressure and will require a pressure cap of 16 to 20 psi. The pressure cap of the expansion tank should be the highest point in the coolant system. The expansion tank is designed so there is space in the tank for the coolant to

Closed cooling systems use an expansion tank with the radiator pressure cap on the top of the tank and not the radiator. A flexible hose connects the overflow tube pipe of the radiator to the expansion tank. This tank should be filled to about one-third with coolant. (Photo Courtesy Jim Halderman)

This modified car uses a Champion remote radiator pressure cap hose filler in a closed cooling system. (Photo Courtesy Champion Cooling Systems)

the atmosphere through the vacuum valve so that the upper radiator hose will not collapse.

There were also early systems that used a recovery or reservoir tank. In this type of system, a hose from the reservoir tank went to a connection just below the radiator pressure cap on top of the radiator. A recovery or reservoir tank uses a vented cap and is not required to be above the cylinder heads. Its job is to hold the coolant that is discharged from the system's pressure relief at the radiator pressure cap when the coolant is hot and expanding. When the system cools, the cooling effect creates a vacuum that pulls this coolant back into the system.

A catch tank was used to collect expelled coolant from the system. You should not confuse the recovery or reservoir tank with a catch or surge tank. The recovery or reservoir tank will either be plumbed at the bottom of the tank or have a hose internally that runs to the bottom so coolant can be drawn back into the cooling

expand. The expansion tank can also be used as a fill point for the system. If an expansion tank is overfilled, it will discharge coolant when the system is at operating temperature.

When an expansion tank is used, the radiator does not use a pressure-relieving cap. The expansion tank's pressure cap acts the same as the regular radiator pressure cap.

When the engine cools down and the coolant in the system also cools, the drop in pressure will open the vacuum valve in the radiator pressure cap. Since the coolant in the reserve tank or expansion tank is vented to the atmosphere, it is now at a greater pressure than the coolant still in the cooling system, so the coolant from the expansion tank will now flow back into the radiator. The radiator pressure cap opens to

This aftermarket GM expansion tank kit can be used for a high-performance cooling system. Some aftermarket firms refer to this as a surge tank, which is defined as a tank connected to a line carrying a coolant and intended to neutralize sudden changes of pressure in the flow by filling when the pressure increases and emptying when it drops. (Photo Courtesy BeCool)

In this 1994 Chevrolet with a closed system, the radiator pressure cap is on the radiator with a hose going from the recovery or reservoir tank to a connection just below the radiator pressure cap. The cap on the recovery tank is vented.

A radiator pressure cap mount has a hose just below the opening going to the plastic reservoir or expansion tank.

system. A catch tank is not under pressure and uses a vented cap. It features plumbing into the tank with a hose that goes back to the expansion tank.

While both a reservoir tank and a catch tank hold excess coolant, a reservoir will pull the coolant back in the system. Whereas a catch tank will hold the coolant until the vacuum valve in the expansion tank opens and coolant is sucked back into the expansion tank when the pressure is low, or in some applications it can be emptied.

Aftermarket Overflow Tank

Higher horsepower engines will produce more heat energy that will need to be dissipated by the engine coolant through the radiator. Generally the best way to improve radiator performance is to make it bigger. Yet you may not have enough space in a modified vehicle for a larger radiator plus the additional weight of the radiator. To assist in this situation, most high-performance radiators are made from lightweight aluminum and come in many different sizes and configurations

The size and the number of cooling fins or fin density and the coolant tube diameter can be manipulated to improve heat dissipation. More fins equals higher heat rejection, so why

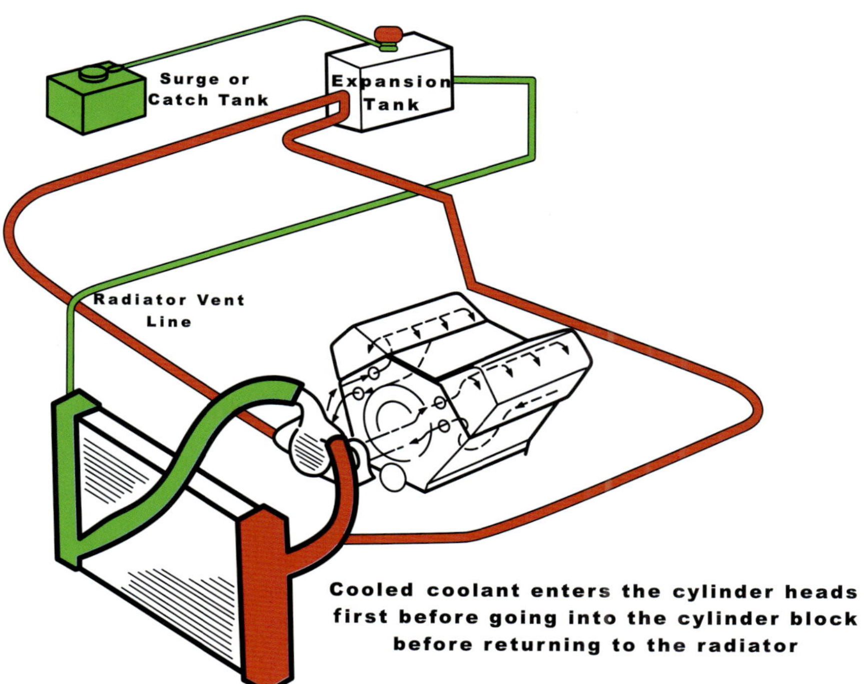

Cooled coolant enters the cylinder heads first before going into the cylinder block before returning to the radiator

From 1992 to 1996, General Motors used a reverse-flow cooling system in the Chevrolet Corvette LT1 engine. This system used a closed cooling system with a pressurized expansion tank along with a nonpressurized surge or catch tank to the left of the radiator. In this system, coolant flows to cool the heads before the cylinder block, which is a reverse of the normal flow.

Champion Cooling Systems offers aftermarket overflow or expansion tanks that can be added to high-performance systems on a modified vehicle. This will give a new system added cooling system capacity. (Photo Courtesy Champion Cooling Systems)

BeCool high-performance downflow radiators in a narrow design are used in older hot rods with narrow space available. Note the mounting brackets and lugs for various electric cooling fan options, as well as radiator mounting points. (Photo Courtesy BeCool)

BeCool offers a large dual-core crossflow aluminum radiator. (Photo Courtesy BeCool)

not make all radiators with the most cooling fins as possible? High fin density is not good for all modified vehicle applications.

The cooling fin count needs to be a balance of the heat dissipation and the amount of area that will be prone to clogging. Clogging reduces flow and cooling and is an issue if a vehicle is being used on a dirt or circle track. If the fin density is too high, it will increase the pressure drop in addition to restricting airflow, causing the pressure drop. This takes place with the more cooling fins in the lesser area for air to pass through and take away the heat.

Heat Rejection Needs

There are many factors in engine combustion and heat transfer that are common, such as engine operating mode, fuel air ratio, volumetric efficiency/mass efficiency, etc. The best approach to radiator sizing is to use actual full-load heat rejection data acquired from the engine original manufacturer. This is usually from dynamometer test data and sometimes can be done using computerized performance simulators. Some manufacturers per-

This is a one-row drag racing radiator. It is a 16x14 overall dual-pass crossflow radiator. (Photo Courtesy Champion Cooling Systems)

Different sizes of radiator are available, such as this one-row 25x13 overall dual-pass crossflow drag racing radiator. (Photo Courtesy Champion Cooling Systems)

form final testing on a test track to determine if the radiator is the right size.

Heat rejection values of an engine within the same family can vary up to 5 percent, such as the GM LS1–LS5 series. So if you are using GM data it's important to get the most accurate information. Something that is just as important is the coolant pump flow rate over a range of engine RPM and temperature, which is generally obtained from the engine manufacturer.

Most OEMs consider this data proprietary and will generally not provide it. Internet searches usually will not yield this data either. The best approach is to write a very detailed letter to the engineering department asking for specific data for the engine you are using. The Chevrolet high-performance engines for the LT1–LT4 and the LS series are used in the Corvette, so you can contact the Corvette Action Center in Louisville, Kentucky, through its website corvetteactioncenter.com. Other aftermarket radiator companies, such as Champion Cooling Systems, BeCool, and U.S. Radiator, have blogs and helplines.

OEM Radiator Design and Testing

When an auto manufacturer designs a radiator, it has to first consider the coolant passage design for the cylinder head. Then, consideration must be given to the engine block, the water pump flow rates, and the thermostat function. All of these things are found when the engine is on an engine dynamometer in the test cell.

Radiator design is based on the amount of heat that needs to be rejected. The design process begins by sizing the radiator to meet the engine heat rejection for any possible operating range. Any internal combustion engine ignites the fuel-air under pressure to produce useful work but also with about 70 percent wasted heat. The wasted heat of combustion is transferred to the coolant through the cylinder wall and the cylinder head water jacket by conduction and convection. The coolant also picks up friction and lubricating oil heat. The heat of the coolant is transferred by forced convection to the atmosphere as the coolant is pumped through the radiator.

Next, designers size the radiator for maximum engine horsepower and load conditions. An engine that produces its maximum horsepower at 6,000 rpm at full load can have a heat rejection into the cooling system of about 6,200 btu per minute. Engine durability or performance are directly related to the coolant temperature, and you want to have a heat output and also want to be able to prevent any possible overheating due a poor design. You would need to design a radiator for the worst-case situation.

Radiator design can start on the dyno, but it must be completed on a test track in the warmer western states. Most OEMs use hot-weather test tracks in Arizona. They can run their vehicles under various load conditions in extremely hot weather, which is a true test of an engine cooling system. They also commonly use Pike's Peak as a test location.

Radiator Selection

When choosing a radiator, you have to consider horsepower, road speed, and loads that the vehicle will be placed under. The airflow through the system, including the air-conditioner condenser, is also a factor. Air flowing through the condenser results in an airflow restriction that must be considered in radiator selection. Other factors include dimensions, fin density, number of passes, number of rows, materials, and fluid.

Dimensions

Radiators used for high performance should be as thin and wide

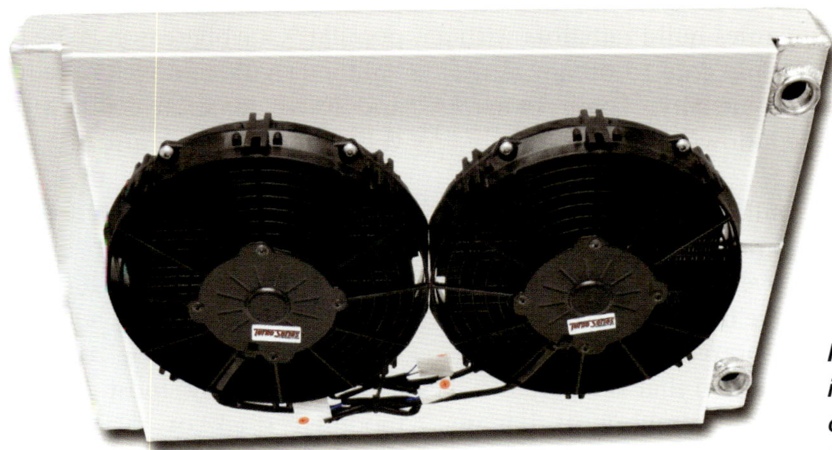

Here is an example of a one-row drag racing radiator with 25x16 overall dual-pass crossflow design. (Photo Courtesy Champion Cooling Systems)

The radiator on the left shows a cutaway of a dual-core design by BeCool. The one on the right shows BeCool's dual-core and dual-pass circle track radiator. (Photo Courtesy BeCool)

as the vehicle will allow. You need a wide and tall radiator with maximum surface area. If the radiator is too thick or has too high a fin density, the air will move too slowly through it to remove the heat from the coolant.

Fin Density

Fin counts are also a critical radiator-design component, but a higher fin density (measured in fins per inch) may make airflow more difficult and not necessarily work well for dirt track or some street applications. As the radiator fin density is increased, the efficiency generally goes down. High fin density provides good cooling, but it can restrict the flow through the engine, causing the pressure to drop. You need to achieve a balance between the number of fins and airflow.

Passes

There are single-, dual-, and triple-pass radiators. All downflow radiators are of a single-pass design, in which coolant moves from the header tank through the tubes to the collector tank at the bottom.

Crossflow radiator design can be single-, dual-, and triple-pass designs. When there are multiple passes through the radiator, the time the coolant is in the radiator increases and results in a lower outlet temperature. However, there is more restricted flow through the radiator, so it is extremely important that you have a high-flow pump in the system, usually somewhere in the 60 to 70 gpm at minimum.

Many of the performance radiator companies build radiators with a dual or triple core to obtain a maximum fin density. These radiators have the following features:

- Billet filler neck that reduces stress cracks or leaks and has large overflow tube
- Effective fin design available
- A CAB-brazed non-epoxy core construction that is repairable
- Aluminum components that are up to 50-percent lighter than copper, brass, or lead radiators
- Internal transmission cooler and upgraded internal engine oil coolers
- Most include OEM design brass drain petcock

Number of Internal Rows

Another factor is the number of rows of coolant tubes. There are two- and three-row designs. Tubes in all radiators are flattened to increase surface area that contacts the fins. Aluminum radiators use 1-inch-diameter tubes that are roughly 31 to 48 inches apart.

American-made radiators use thicker wall tubes that are less likely to fail under high pressure. Then why have aluminum radiators become so popular? One big reason is the potential for a significant weight reduction and lower material costs. Modified vehicle builders are also big on aluminum radiators for that reason, with a weight difference of around 10 to 15 pounds.

Radiator Materials

Aluminum is the material of choice today by most builders of modified stock or racing vehicles. It is lighter, stronger, and can have multiple pass designs. They also have welded tanks, which generally eliminates leaks and air ingress.

In the past, copper and brass radiators were used for good heat transfer rates. Copper has an excellent thermal conductivity rating. A copper fin's thermal conductivity rating is about 50 percent higher than an aluminum fin. Brass, which is an alloy of copper, is not as good of a conductor as aluminum but is used for the tubes because of its strength. One problem with copper is that the lead solder used in older copper and brass radiators has a terrible thermal conductivity rating, which limits the efficiency of lead-soldered radiators. So some companies, such as U.S. Radiator, have instituted a newer process that improves efficiency by changing the flux and solder and its contact with the fins.

Some copper and brass radiators are less expensive than others due to their construction. The original radiators built in the muscle car era for GTC, Olds 442, Road Runner, etc. used 11/42-inch tubes. More modern radiator construction moved those centers closer together, with the same 11/42-inch tubes. This creates room for more tubes in the same-size radiator core. Each of these versions can be obtained in two-, three-, or four-row applications. As the radiators become denser, they become more expensive. However, the thicker the radiator, the more airflow will be reduced.

Aluminum radiators can be expensive, costing between $450 and $550. Yet, there are universal crossflow aluminum radiators with no mounting tabs that have either General Motors or Ford-style inlet and outlet hose configurations. These radiators are a two-row design with 1-inch tubes and come with

The BeCool downflow radiator shown here has an OEM-style automatic transmission fluid cooler located in the bottom or outlet tank of the radiators. This is the coolest part of the radiator and the best location for an internal fluid cooler. Note the brass fittings to the left of the petcock. (Photo Courtesy BeCool)

a machined-aluminum filler neck welded into place. If you are willing to do some radiator mount fabrication, you can fit these radiators to many different applications.

Some of aftermarket radiators do not have an internal automatic transmission fluid cooler, so you may need an external transmission fluid cooler. When you use an external cooler, you may experience some high transmission fluid temperature in heavy traffic, so you may need to install a fan on your external transmission fluid cooler. You may also have to fabricate a shroud for this universal crossflow radiator to optimize airflow through the radiator.

Automatic Transmission Fluid Coolers

All OEM-built vehicles supply automatic transmission–equipped vehicles with an automatic transmission fluid (ATF) cooler installed in one of the radiator tanks. That is about 90 percent of production vehicles. Most of the high-performance radiators come equipped with an ATF cooler in the left tank.

The automatic transmission works harder when it is used for racing, it has added weight from a large-displacement engine, or it is used for towing. The transmission can get hotter, and heat is one of the major enemies of ATF. An aftermarket transmission cooler can keep your transmission from getting too hot, helping you get the best performance and long life out of it. Some drag racing radiators do not have a transmission fluid cooler. If you are

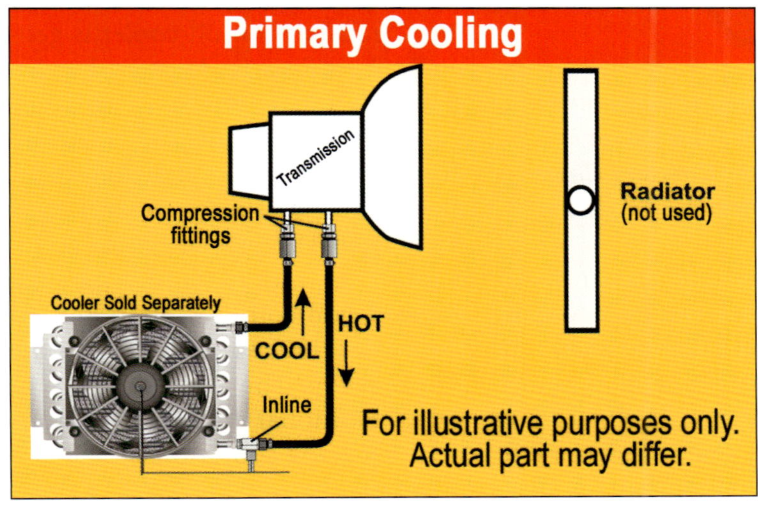

A remote automatic transmission fluid cooler is installed away from the cooling system radiator. (Photo Courtesy Derale Performance Inc.)

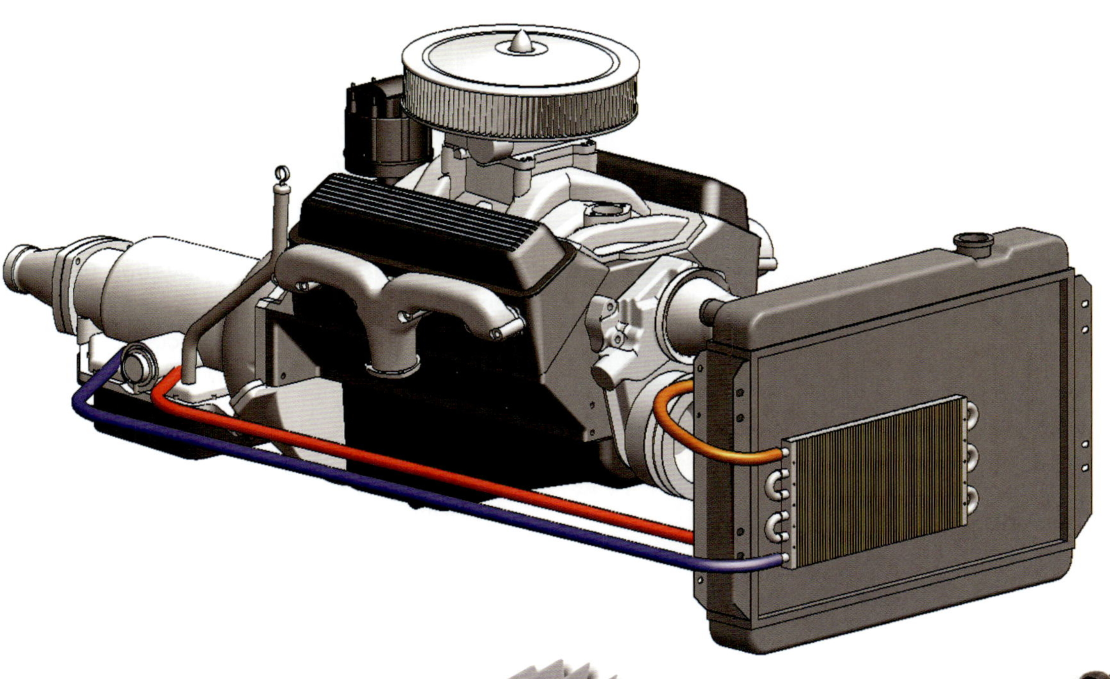

For a Derale auxiliary automatic transmission fluid cooler setup for a modified vehicle, the cooler is placed in front of the radiator with inlet and outlet lines running to the automatic transmission. (Photo Courtesy Derale Performance Inc.)

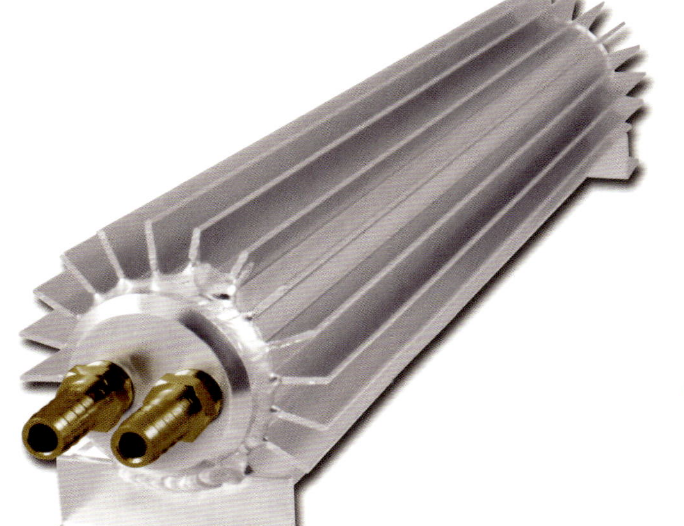

A Champion cooling external automatic transmission fluid cooler is shown here. (Photo Courtesy Champion Cooling Systems)

A remote six-pass auxiliary automatic transmission fluid cooler does not require placement in front of the radiator. (Photo Courtesy Derale Performance Inc.)

running an automatic transmission that does not have one, you would need to add an auxiliary cooler.

When a transmission fluid cooler is used in the radiator, it is placed in the outlet tank, where the coolant has the lowest temperature. The auxiliary transmission oil cooler is generally an oil-to-air heat exchanger placed in front of the air-conditioner condenser or radiator. The transmission oil temperature is regulated by the airflow passing over this heat exchanger. The oil out of the transmission is plumbed through the transmission oil cooler hoses to the cooler then directed back to the transmission. This cooler helps provided additional cooling for performance driving conditions.

Some project vehicles cannot have the ATF auxiliary cooler in front of the engine cooling system radiator. If that is the case, there are three options: a remote mounted auxiliary cooler with its own cooling fan, a heat sink–style cooler, or a frame rail–mounted cooler.

Selection Example

When selecting a radiator, it should be as thin as possible and have the least number of fins per inch. You would most likely select a 0.7-inch-deep single-row radiator for

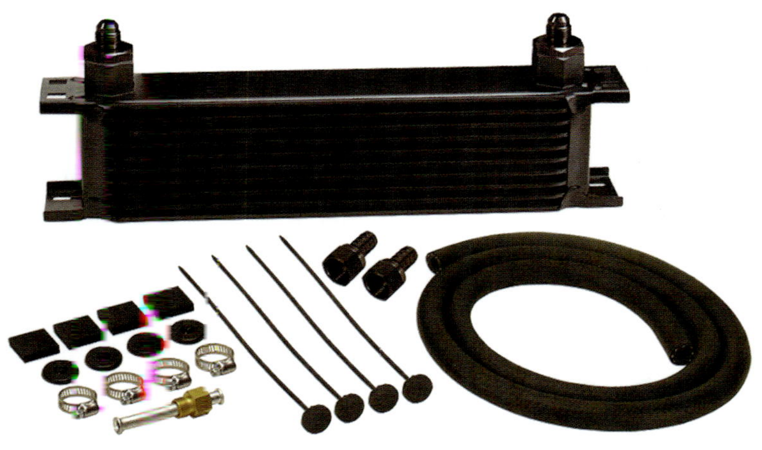

Another option is a frame rail–installed auxiliary transmission fluid cooler. (Photo Courtesy Derale Performance Inc.)

your first choice to establish a performance curve.

Let's say you are selecting a radiator for a high-horsepower crate engine such as a GM LS1–LS5 series or the new LT1 crate engine to be installed in a modified street vehicle, I would suggest the following:

- Select a heavy-duty radiator size that the OEM used with this engine in a production vehicle, such as the Cadillac CT5 or Corvette new LT1.
- Choose an aftermarket welded–aluminum dual-pass radiator that will fit along with dual electric fans with proper shrouding or a tri-flow triple-pass radiator with proper shrouding. You can consult with any of the radiator companies for selection advice.
- Install a high-capacity water pump (either mechanical or electric).

Radiator Inefficiency

Radiators can become inefficient or defective. This, in most cases, will lead to engine overheating. Let's look at a couple of the more common problems.

Partially Blocked Radiator Core

If the outside surface of the radiator core is blocked by road dirt and debris, you will have much less air flowing across the core. This will mean the heat absorption will be low, which could result in overheating. You can clean out airways by applying air or water pressure from the engine side of the radiator. This will reverse the direction of normal airflow.

Partially Blocked Waterways

Corrosion on the internal metal surfaces of the cooling water jackets can act as a thermal insulator. This slows down or prevents the conduction of heat through the metal. The remedy is to reverse flush the radiator.

A matrix flow test will determine the extent of any internal restriction to coolant flow to the radiator. To perform the test, remove the hoses and temporarily insert a plug in the outlet pipe. Fill the radiator and container completely with water then remove the plug from the outlet pipe. Record the time it takes for the water and container to drain and compare it to OEM specifications.

Standard Cooling System Flushing Steps

1. Drain the system (dispose of the old coolant correctly).
2. Fill the system with clean water and flushing/cleaning chemical.
3. Start the engine and run until it reaches operating temperature with the heater on.
4. Drain the system and fill with clean water.
5. Repeat until the drain water runs clear (any remaining flush agent will upset pH).
6. Fill the system with 50-50 antifreeze/water mix or a premixed coolant.
7. Start the engine and run until it reaches operating temperature with the heater on.
8. Adjust coolant level as needed.

Reverse Flushing a Cooling System Steps

1. Flush the water through the radiator until it runs out clear.
2. If water does not clear in a few minutes, remove the radiator and turn it upside-down or stand it on one end so that you can flush it in the reverse direction to the normal flow. Tilt if necessary to bring the inlet at the top or header tank to the lowest point.
3. Use a garden hose into the bottom or collector tank to push water out the top or header tank of the radiator. ∎

COOLANT FLOW

Water-cooled engines use a centrifugal pump to move coolant through the cooling system. The water pump got its name from early internal combustion engines using water as the coolant and pumped through the system.

The first machine that could be considered a centrifugal pump was a mud-lifting machine that appeared as early as 1475 in a treatise by the Italian Renaissance engineer Francesco Maurizio di Giorgio Martini. A real centrifugal pump was not developed until the late 17th century, when Denis Papin built one using straight vanes. The curved vane was introduced by British inventor John Appold in 1851.

A centrifugal pump works much like when a person holds a rope with a pail of water attached to it and swings it around his or her head. No water is spilled as long as the pail is moving fast in a circular path. Centripetal force is the inward force pulling down on the rope held by the person's hand, which keeps the pail and the water from flying out of the hand. The reactionary centrifugal force holds the water against the bottom of the pail and is opposite to the centrifugal force the water exerts on the bottom of the pail.

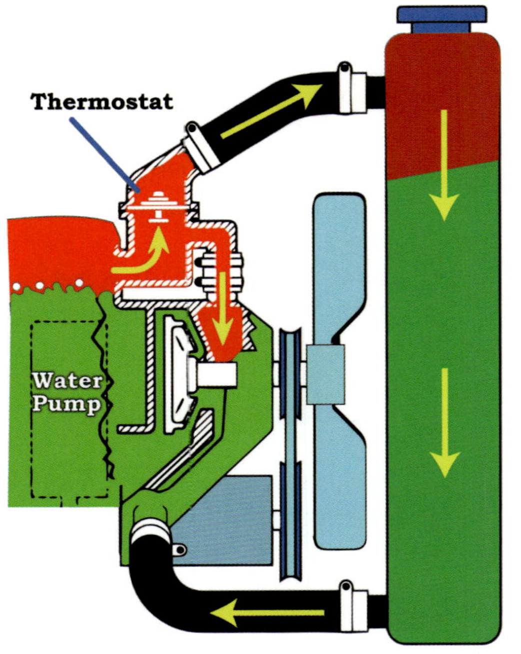

Water-cooled engines use a volute or diffuser-type circulation pump to pump the engine coolant through the cooling system. The crankshaft pulley runs through a drive belt and turns the water pump, pumping low-temperature coolant into the engine and carrying away the heat of combustion. It pushes the coolant out an open thermostat and back into the radiator where the heat will be removed via convection.

Now drill a hole in the bottom of the pail and attach a curved outlet pipe into the hole. Then assume that the supply line feeding the water into the pail rotates in the same direction as the person's arm. As the pail swings, water is forced through the hole and discharged through the curved pipe. As the water pail moves at an increased speed along its circular path, the water will discharge to a greater extent from the curved pipe.

The hole at the bottom of the pail represents the pump discharge nozzle or pump outlet. The supply line going into the top of the pail is equivalent to the suction line or pump inlet of the water pump. The water pail represents the water pump housing or body, and the person's arm and rope together equal the pump impeller.

As the impeller is turned by the engine (arm swings the pail), coolant will be forced out of the pump discharge by the centrifugal force due to the rotating speed. The energy transferred to the coolant by the centrifugal force is converted to pressure. The centrifugal water pump is a self-priming pump because the impeller and diffuser (if used) are always submerged in coolant.

Centrifugal Water Pump Components

Centrifugal pumps move engine coolant by the conversion of rotational kinetic energy to the hydrodynamic energy of the fluid flow. The engine crankshaft pulley or camshaft via idler gear provides the rotational energy. Coolant goes into the pump impeller near the rotating axis and is accelerated by the impeller, flowing radially outward into a diffuser or volute chamber, where it exits into the engine water jackets or cylinder heads in a reverse-flow cooling system.

This is a cutaway of a 4-cylinder engine using a vane-type impeller. The water pump uses centrifugal force to circulate the coolant. It consists of a fan-shaped impeller set in a round chamber called a volute with curved inlet and outlet passages. (Photo Courtesy Consulab Corporation)

Impeller

The impeller is rotated by the crankshaft or camshaft. In a centrifugal pump, the coolant is forced by either atmospheric or system pressure into a set of rotating vanes. This establishes an impeller that discharges the coolant at a higher pressure and velocity at its edge.

Vanes

The diffuser vanes change the direction of the coolant flow and convert the velocity energy into pressure energy. The impeller can have straight (open) vanes, ones shaped like a spiral scroll (also called a

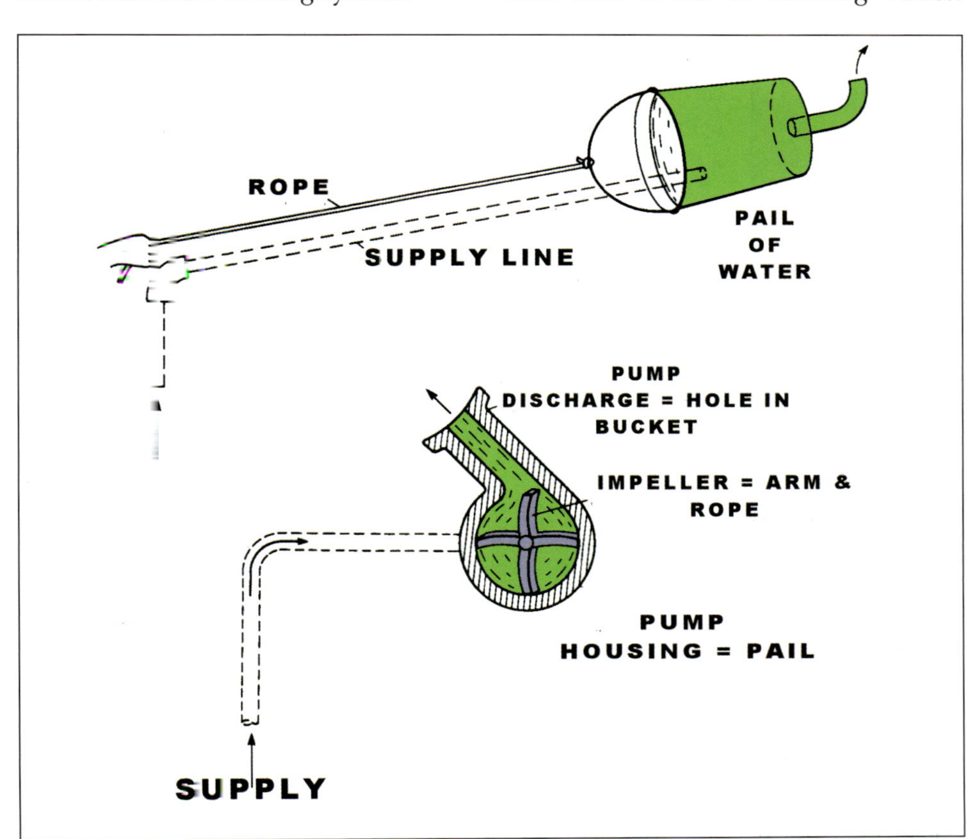

Swinging a pail of water around your head simulates the operation of a centrifugal water pump.

This centrifugal water pump drive hub shaft, bearing, and scroll impeller is shown before the bearing and hub shaft are installed into the pump housing of an Edelbrock 8810 high-flow water pump. (Photo Courtesy Edelbrock Corporation)

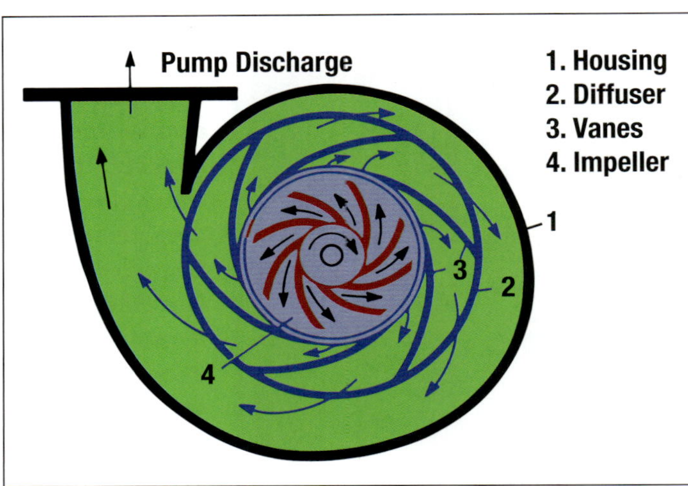

A diffuser-type centrifugal water pump uses a diffuser or dividing wall that distributes the radial forces created by the rotating impeller to change the direction of the coolant flow.

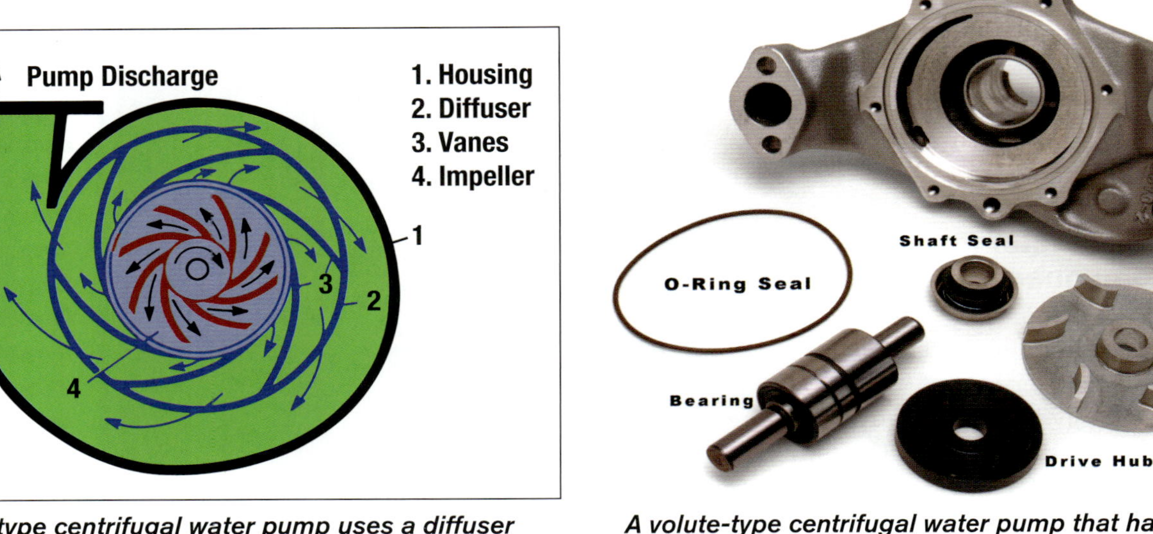

A volute-type centrifugal water pump that has been dis-assembled shows the volute housing, the bearing, seals, hub, and scroll-type impeller. (Photo Courtesy Edelbrock Corporation)

This cutaway illustrates the inner workings of a diffuser-type pump with a semi-open scroll impeller. A scroll-shaped chamber with a straight vane impeller are used in some water pumps. (Photo Courtesy Jim Halderman)

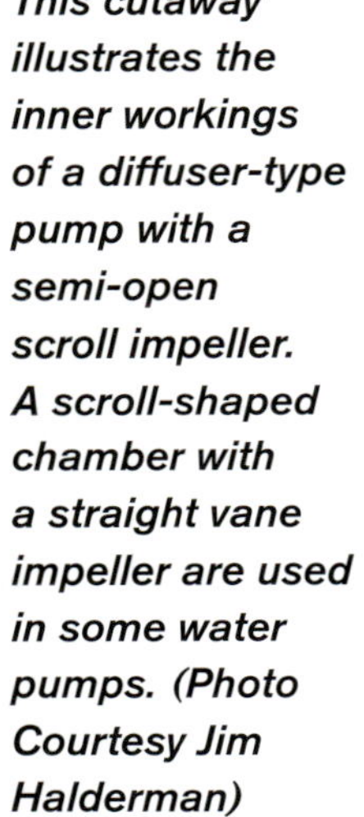

There are also sheet metal–type open-design impellers that can be used on 4-cylinder engines. They are manufac-tured from stamped steel as opposed to the more expensive

forged-aluminum designs. (Photo Courtesy Jim Halderman)

semi-open impeller), or closed vanes. Spiral scroll–shaped impellers with a diffuser are characteristic of high-flow pumps. Some water pumps use a scroll-shaped passage with a vane-type impeller. Some impellers in OEM pumps are made out of stamped steel.

Volute

The typical centrifugal water pump housing will generally have a doughnut-shaped volute. A volute is a spiral casting designed so pump speed will be converted to pressure without any shock. Some will have diffuser-type vanes because the housing will contain a set of stationary diffuser vanes surrounding the pump impeller.

Centrifugal Water Pump Operation

The term *centrifugal* is defined as acting or moving in a direction away from the axis of rotation or the center of a circle along which a body is moving. Centrifugal force can be

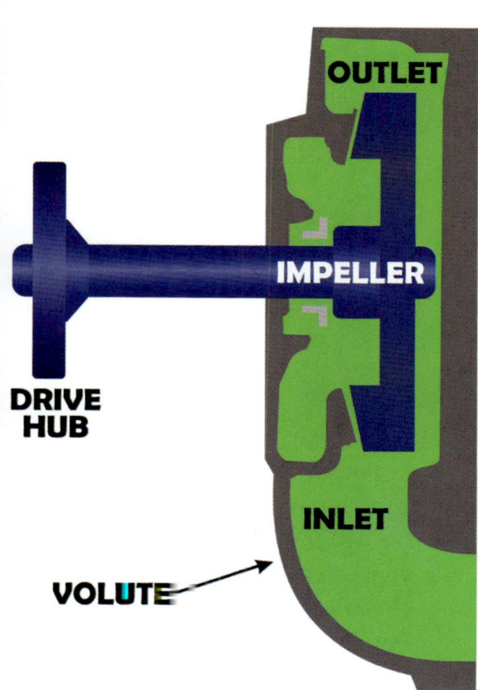

The crankshaft-driven hub drives the water pump impeller inside the volute to move engine coolant by accelerating it outward using the impeller toward the volute casting. A volute is a spiral casting designed so that pump speed will be converted to pressure without any shock. (Photo Courtesy Mascot Graphics Inc.)

The impeller can also be shaped like a scroll, which is a spiral design that provides a lot of turbulence. This type of impeller is referred to as a semi-open impeller in engineering terms. (Photo Courtesy Meziere Enterprises)

experienced when a stone is whirling around on the end of a string in a flat plane. The only real force acting on the stone in the flat plane as applied by the string (gravity acts vertically downward) is called centripetal force. Centrifugal force is the experienced force not the actual force, where centripetal force is the real vector force.

A centrifugal pump is a machine that moves engine coolant by accelerating it outward using an impeller toward a volute casting or vanes. A centrifugal pump converts rotational energy from an internal combustion engine through the drive belt or gear to the pump hub into energy in the moving coolant. Centrifugal pumps with volute castings are called volute pumps, and those with diffuser vanes, such as a turbocharger, are called diffuser pumps. Diffuser pumps contain a set of stationary diffuser vanes surrounding the pump impeller.

A portion of the energy goes into kinetic energy of the coolant. Coolant enters axially through the eye of the casing, is caught in the impeller blades, and twirls tangentially and radially outward until it leaves through all circumferential parts of the impeller into the volute or diffuser part of the casing. The coolant gains both velocity and pressure while passing through the impeller.

The doughnut-shaped volute or diffuser (sometimes called a scroll when it is spiral shaped) section of the casing decelerates the flow and further increases the pressure. The coolant is not pushed radially outward by centrifugal force (nonexistent force), but rather by inertia, the natural tendency of an object to continue in a straight line (tangent to radius) when traveling around circle. You could compare this action to the way a spin-cycle works in a clothes washer.

The inlet to the water pump from the collector side of the radiator (cooled coolant) is located near the center so that coolant is sucked out of the radiator by the pump vanes. Using centrifugal force, the pump vanes fling the coolant to the outside of the pump, where it enters the engine. The coolant leaving the pump flows first through the engine block and cylinder head, returns through the thermostat and into the radiator, and finally back to the pump.

The engine block and cylinder head have many passageways cast or machined in them to allow for coolant flow. These passageways direct the coolant to the most critical areas of the engine.

The water pump is driven by a belt or a gear connected to the crankshaft, or in the case of the reverse-flow GM LT1, it is directly connected to the camshaft. The pump circulates fluid whenever the engine is running. The water pump uses centrifugal force to pump coolant to the outside while it spins, causing fluid to be drawn from the center continuously using centripetal/centrifugal force.

On an engine dynamometer, it will show the amount of coolant flow that actually occurs through the cooling system. The typical discharge rate is about 70 gpm with a cooling system capacity of 6 gallons with about 3 gallons in the engine.

The GM LT1 reverse-flow cooling system was used primarily in the Chevrolet Corvette from 1992 to 1996. This water pump is driven directly by the engine camshaft through splined shaft from the camshaft sprocket spur gear to the water pump impeller. This pump operates at a 1:2 ratio, so the pump is turning at twice the speed of the crankshaft. This water pump draws coolant from the radiator by the water pump and passes the coolant through the thermostat on the inlet side of the pump. The coolant is routed through the cylinder heads first before the cylinder block water jackets. (Photo Courtesy Jim Halderman)

The water pump on this overhead camshaft 4-cylinder is driven by the cog-style timing belt. (Photo Courtesy Jim Halderman)

The water pump on a dual overhead camshaft (DOHC) engine is driven directly by one of the camshafts. (Photo Courtesy Jim Halderman)

On an engine dyno stand, a high-flow water pump shows the amount of coolant or water that is pumped by a high-volume flow water pump. (Photo Courtesy Jim Halderman)

Hydraulics

Engine coolant needs to be moved through the engine block water jackets and cylinder heads to remove the heat of combustion. It is then moved to the radiator so that that heat can be removed through convection. An efficient cooling system requires that the coolant has to be kept in motion otherwise it will eventually boil. This is done through the hydraulics, which is the transmission of liquids such as engine coolant through hoses and channels that can be a source of mechanical force or control.

Pressure-Flow Relationship

Water pump hydraulics have a pressure-flow relationship. The rule is the coolant flow is inversely proportional to the pressure, so as the flow decreases the pressure increases or as the pressure increases the flow decreases. Coolant pressure increases as it expands and absorbs heat from the energy required to pump it through the engine and into the radiator.

Closed cooling systems use a nonvented radiator to help build pressure in the system (discussed in chapter 3). The engine cooling system has passages in the radiator, engine block, water jackets, and cylinder head along with the thermostat hoses and heater core that coolant must flow through. These passageways are constantly changing in dimension and shape during the flow to fit the space available or to route the coolant where it needs to go in the system.

The engine coolant is forced through an orifice, such as a thermostat, which increases system pressure. So the design has to consider these various pressure fluctuations for the benefit of the cooling system. Water pump flow is also affected by the resistance or head pressure that it must work against. The pump flow resistance is the spreading of the coolant pressure drop through the radiator, coolant hoses, and the engine itself. As a hedge against this head pressure, the amount of power required to turn the pump impeller will also increase.

Cooling System Flow Direction

Coolant as a hydraulic fluid does not like to change its travel direction. Every time the coolant is required to make a turn, the flow and the pressure drop before the turn increases. One could say it's not possible to design a cooling system that has no turns. A decrease in cooling system pressure can be controlled by decreasing the angle of the turn. A properly designed water pump should supply acquired coolant flow to the radiator, providing the plumbing contribution to the total system restriction is

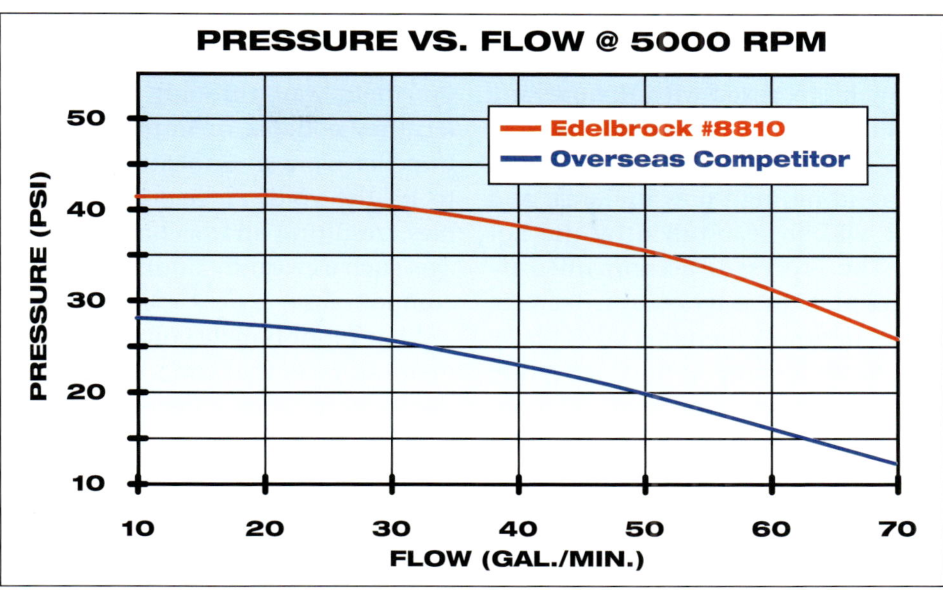

This graph from Edelbrock shows the pressure versus flow performance of its Victor 8810 high-flow water pump. As the flow decreases, the pressure increases. (Diagram Courtesy Edelbrock Corporation)

able to be overcome. For this reason, your cooling system design needs to take a systems approach and cannot be a one-dimensional solution, such as just a big radiator.

All centrifugal water pumps have separate inlet and outlet sides. They are referred to as the suction or inlet side and the pressure or outlet side. There are two basic requirements that are needed to maintain good fluid movement in any cooling system. The first is the pump's ability to generate sufficient pressure to overcome the head pressure and system resistance at the outlet. This requirement does not present any hydraulic limits because if it has sufficient pump sizing and rotation speed, almost any flow level can take place.

The second pump requirement relates to the suction side of the water pump, where a pressure rise is needed between the impeller and the suction pressure to draw the coolant into the pump. The amount of low pressure at the leading edge of the impeller vanes will vary the flow. The lower the pressure, the greater the flow. However, you can go too far. If the coolant in the pathway reaches vapor pressure, vaporization and cavitation can take place. This will lead to loss of prime and erosion.

Water Pump Design Issues

There are a number of issues that you will need to consider regarding the water pump: air, cavitation erosion, and pump speed. Air can be a big problem when it gets into your cooling system. It is important to have a tight and fully sealed system at the gasket and hose clamping areas. Water pump speed and cavitation areas are equally important.

Air

Air is the biggest enemy of an internal combustion engine cooling system. The presence of vapor at any point in the coolant flow could cause flow and pressure variability not only

---

at the pump source but also over the entire pressure field as the vapor bubbles will be carried with the pressure stream.

Some bubbles will collapse in the surrounding high-pressure area, and some will be reabsorbed by the liquid. The process is accompanied by strong pressure waves that will create surges in flow. When this takes place, significant energy is lost, affecting pump efficiency. You can tell your system is air-bound if you have a temperature gauge that swings from hot to normal temperature.

Cavitation

Cavitation is the formation of vapor cavities in a liquid, which occurs when the pressure within the water pump drops. These cavities (or bubbles) are the result of forces acting upon the coolant when it is subjected to rapid changes of pressure. Low pressure can be caused by various flow parameters, such as coolant viscosity, temperature, pressure, and flow nature.

When these vapor-filled bubbles get to areas of higher pressure on their way through the impeller, they collapse or implode on the impeller vanes. Coolant rushes in to fill the voids created by the bubbles, resulting in mechanical damage, such as wear in aluminum pump components.

If a vapor bubble collapses near a metal surface, that surface may deteriorate very rapidly. Repeated implosion can result in surface fatigue of the metal, causing a type of wear also called cavitation. The most common examples of this kind of wear are to pump impellers and bends in the volute and coolant passages where a sudden change in the direction of liquid occurs.

Water pump efficiency is also affected by pressure. For example, if no radiator pressure cap is used in the system, a water pump is only about 85-percent efficient. Using a 14-psi cap, the pump becomes almost 100-percent efficient because the pressure decreases cavitation.

A water pump that has been heavily damaged by cavitation erosion and corrosion can be seen here. It clearly needs to be replaced. (Photo Courtesy Jim Halderman)

A pressurized system makes it difficult for the vapor bubbles to form. Unchecked, cavitation will erode the water pump blades and housing

Water Pump Speed

If all of the parameters of the pump design are known, then it is mathematically possible to calculate the maximum operating speed of the impeller. The specific pump rotational speed is an index speed used to determine the profile or shape of the impeller that is most suitable near the point of best pump efficiency. Specific pump speed is defined as the speed (RPM) at which an impeller will operate when reduced in diameter to produce a flow of one gallon per minute (GPM) against a total pump head of one foot.

The theory of specific pump speed is associated with the upper operating limits of head and capacity that can safely be handled by a specific pump without cavitation. (Head is defined as a specific measurement of liquid pressure above a specific point. A column of water in a vertical pipe exerts a certain pressure on

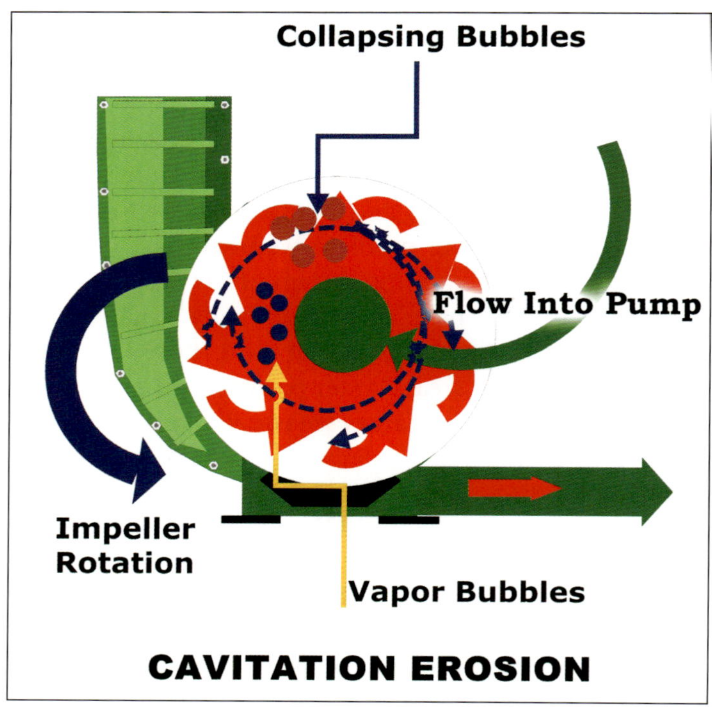

This illustration shows how cavitation erosion can take place after vapor bubbles (shown as blue dots) collapse when the pressure is low in the volute of the water pump.

the horizontal surface at the bottom of the pipe expressed in psi or feet of liquid. The height of this column is called head).

All pump characteristics are dependent on the specific speed; therefore, this measurement is very important to water pump design. Water pump manufacturers generally have the accumulated data to help with your cooling system design.

Engine Coolant Flow

Most automotive text books will tell you to never leave out the thermostat. This came from when early Cadillac Motor Car Division V-8 engines overheated when the thermostat was removed. This led to the thinking that if the coolant travels through the radiator too fast, it would not have enough residual time in the radiator to remove the heat. This issue became firmly entrenched in the modified street car as well as the racing community, so it is common to place a restrictor of the cooling system to slow down the coolant flow.

The thinking was that your race car or modified engine would have a lower temperature because the coolant would stay in the radiator longer. When you slow down the coolant through the use of a restrictor (thermostat), it does not improve the engine's ability to remove heat; rather it impedes it by reducing high-velocity coolant flow.

With a coolant restrictor in place, the system pressure can build in the volute area so that the boiling point of the liquid is elevated. Tests conducted by water pump manufacturers on water pump flow test benches have recorded localized pressures as high as 60 psi around the water jacket of a 350-ci Chevro-

let small-block engine with a restrictor at about 6,000 rpm. The high boiling point would allow more heat to be transferred to the coolant and would show a lower temperature reading.

This may sound good, but it is restricting the flow rate for high pressure instead of moving the proper amount of coolant through the system. Coolant will give off more heat when going through the radiator at a high flow rate due to higher levels of turbulence. The faster the coolant moves, the more heat is removed from the engine coolant and the temperature will be lower.

If the restrictor theory were true and slower was better, then why does the industry sell high-flow water pumps? High-performance cooling system component manufacturers do not sell slow-flow systems. However, thousands of restrictors are sold every year to rodders who also invested in high-flow water pumps.

Flow Rates

As stated previously, water pump speed is rated in gallons per minute (GPM). It needs to be established at a specified head pressure. A good factory water pump puts out a minimum of 100 gpm at its maximum engine speed. Water pump manufacturers, such as Edelbrock, have used a low bench and have come up with applications as low as 60 gpm, while others are as high as 130 gpm. Open port flow is meaningless when it comes to engine water pump because it's not against any resistance.

Water Pump Construction

Water pump housings (including the volute and impeller) are made from diecast aluminum, and the shafts and bearings are made from steel. The housing has inlet and outlet suction and pressure sides along with the volute and a bore for the shaft-sealing bearing.

A vent or weep hole was usually cast into the snout shaft bearing boss. The purpose of the weep hole was to allow easy installation of the bearing and seal. If there was no weep hole, the bearing and seal could not be pressed into the casting of the housing because there would be no way for the air to escape.

The impeller is generally press-fit onto the shaft. In addition, there is generally a pulley pressed onto the front of the shaft so you can attach a belt or gear drive to it.

A volute-style water pump housing contains a doughnut-shaped volute chamber with a bore in the casting for the bearing. (Photo Courtesy Edelbrock)

This Edelbrock Model 8810 high-flow and high-pressure water pump shows the weep hole located just below the bearing boss. The weep hole was not placed to allow coolant to weep out or to vent the system; instead, it was to allow an easy installation of the bearing and seal. (Photo Courtesy Edelbrock)

This is a scroll chamber of a water pump with a closed water wheel–style vane. The impeller worked well at low speeds but was generally deficient at high speeds. (Photo Courtesy Jim Halderman)

Shown here are two water pumps with semi-open scroll design impellers. The impeller on the left is made from forged aluminum, and the impeller on the right is made from stamped steel.

The shape and style of the impeller water pump and its blades define the character of the water pump. Energy transfer from the impeller shaft to the coolant is accomplished by forcing the liquid within the impeller to rotate. The torque applied to the impeller will be transformed into centrifugal forces acting on the liquid particles, which in turn will translate themselves into pressure and motion.

High-Performance Water Pumps

There are several factors to consider when choosing a water pump for a modified application. First, the pump impeller is always a compromise because no impeller can be 100-percent efficient at every operating speed. For this reason, it is best to match the design of the water pump with a pulley ratio to create a high-flow coolant system that uses as little power as possible.

Pulley ratios are very important because the pump should run at a speed above engine RPM. Back in the muscle car days, General Motors put big engines such as the 396 and 427 ci in its muscle cars and used a 1:1 pulley ratio. This was done because these engine would run at a high RPM and needed to keep the coolant pump near the peak of its flow efficiency during the operating speed. However, it was common for these engines to overheat when they were driven at lower speeds, such as in traffic. You need to drive the impeller at high speeds to achieve the best coolant flow.

General Motors designed and built a high-pressure and -volume water pump for its LS1-LS5-series engines, which is now in generation 5. This engine is currently used in Corvettes and Camaros as well as light-duty trucks and the Cadillac CT5. The Cadillac CT5 develops more than 600 hp. General Motors designed this water pump for speeds

The Edelbrock version of 1997–2004 GM LS1- to LS6-series high-performance V-8 water pump is a high-volume and high-performance enhancement of the original water pump. (Photo Courtesy Edelbrock)

Edelbrock's high-volume and -performance 8896 water pump for 1997–2004 GM LS1- to LS6-series high-performance V-8 engines is shown. Also available is the Edelbrock 8893, a high-performance street, heavy-duty aluminum water pump for 2005–2008 GM LS cars, and model 8894 for 2009–2016 GM LS cars. These pumps provide increased water flow and pressure, ultimately improving engine cooling system efficiency. They are compatible with all OEM accessories, and installation is the same as for original equipment water pumps. Consult the online service information for specific procedures. When installing the new water pump, tighten the water pump bolts a first pass to 11 ft-lbs. Then, tighten the water pump bolts a final pass to 22 ft-lbs. (Photo Courtesy Edelbrock)

below 4,500 rpm, which is below the approximate 6,000 rpm redline for the engine.

The stock OEM pump works well for most street use, but if an application is for higher RPM racing applications, there may be issues. When above 4,500 rpm, there can be piston ring deterioration because the pump was not designed to move large amounts of coolant above 4,500 rpm. However, increased water-flow and high-pressure output versions of these OEM pumps are offered by Edelbrock and Holley. These pumps are designed to run efficiently up to 10,000 rpm and move coolant fast enough without fear of cavitation.

Serious racing or modified stock vehicles need a high-performance water pump. There are high-flow/pressure water pumps made by Edelbrock and Holley that will meet the bill. Electric pumps are increasingly viable due to their ability to be connected to the CAN-BUS in newer electronic control system vehicles.

Pump Pulley Ratios

Standard passenger cars have a crankshaft to pulley ratio of about 1:1.25 to 1:1.33, and the average production vehicle spends most of its time at the lower end of the engine RPM range. Back in the muscle car days of the 1960s, it was common to find a crankshaft to pulley ratio of 1:1 on the big-block engine used in Corvettes, Chevelles, Hemis, and Fords. The engineers did this because they thought these engines would run constantly at high RPM and they wanted to keep the water pump near its peak flow efficiency. However, it was common for these engines to overheat in traffic at lower speeds.

Back in those days, I was a GM service rep. We would replace radiators and water pumps to try and fix the overheating. Finally, the factory came out with a new crankshaft pulley.

Quality high-performance pumps can run at 4,500 to 6,500 pump rpm on a high-performance street or strip application, but a dedicated race car will usually need a pulley ratio at 75 percent of the maximum crankshaft spins. Most high-performance pump companies recommend a crankshaft to water pump pulley ratio of 1:1.4 to 1:1.5 for street engines that traditionally run primarily at low engine speeds. This ratio would be a good compromise for your street rod.

Electric Water Pumps

The mechanical water pump has been considered by some as an inefficient component designed as an accessory for the early combustion engine. Electric pumps are preferred by drag racers because they eliminate the belt drive and permit liquid circulation for cooldown between runs. They are generally more efficient than engine-driven pumps as well.

The belt-driven mechanical water pump runs at the same speed as the engine regardless of the engine temperature. With an electric water pump with a digital controller, the electric water pump speed can be controlled by the controller, which varies the supply voltage to the pump using pulse width modulation (PWM) to vary the pump speed up or down to achieve a specific engine temperature. When the engine reaches this temperature, the controller changes the pump speed to account for whatever conditions are present while preserving the set temperature independent of the engine speed. The electric water pump reduces the parasitic losses taken by the use of an engine drive mechanical water pump.

There has been a misconception for about 20 years that electric water pumps are primarily used by NHRA drag racers due to their ability to circulate coolant between runs when the engine is not running, as well as the elimination of the belt drive. In 1997, Meziere Enterprises introduced the first electric pump of its kind, using the same seal technology found in factory and OEM pumps. This move dramatically increased the reliability compared to previous versions that had lip seals that only lasted 400 to 700 hours.

The carbon-ceramic seal technology proved to be very reliable. When tested to fail in the lab, the brushes and commutator would wear out before the seal. The expected life on a standard Meziere Enterprises unit is 2,400 hours, and on the heavy-duty or high-flow units the life is 3,000 hours.

Meziere also started designing street style pumps. Many of these designs (GM LT/LS, Ford Modular, and Coyote, etc.) incorporate an idler around the electric motor. This idler takes the place of the stock pulley so that belt routing is not complicated when moving from a traditional mechanical pump to an electric one. These cooling system setups typically gain 8 to 15 hp and better low-speed cooling.

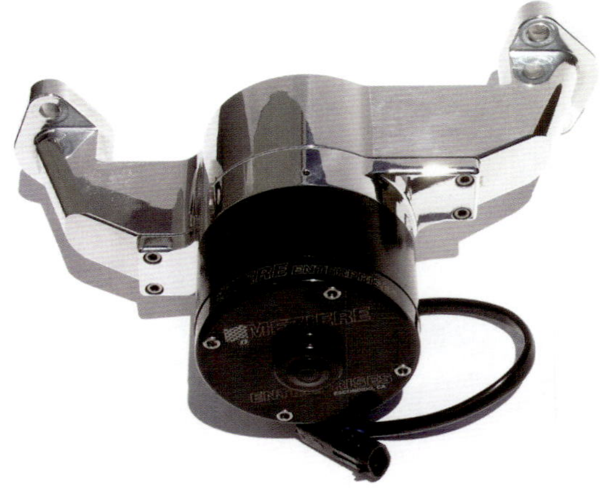

The Meziere 300 series high-flow electric water pump has an impeller and pump cavity that allows for greater volume of water. The heavy-duty motors provide increased torque and RPM. The resulting flow rate of 55 gpm is enough to cool anything from a 600-hp circle track car to a 2,200-hp Pro Mod. This unit is also recommended for supercharged, nitrous-oxide, and high-performance street engines. (Photo Courtesy Meziere Enterprises)

Meziere Enterprises offers a street version for the GM LS1-LS5-series engine. Its electric water pump has a 55 gpm flow rate and easy installation because it accommodates the factory accessory belt. It can free up more than 11 rear wheel horsepower in most applications. (Photo Courtesy Meziere Enterprises)

Electric Water Pump Manufacturers

There are several manufacturers of electric water pumps.

- Allstar Performance Remote Mount Electric Water Pumps
- CSR Billet Pumps
- CVR Pumps
- Davies Craig Pumps
- Dedenbear Remote Water Pumps WP3
- EMP Stewart Pumps
- Meziere Enterprises
- Moroso Billet Aluminum OEM Mount Electric Water Pumps
- Proform Pumps
- Summit Racing ■

High performance meets street practicability in Meziere Enterprises' high-flow 55-gpm water pump for Chevrolet engines with a heater or bypass port. Fittings are available for a wide variety of hose connections. You can also connect the heater core. (Photo Courtesy Meziere Enterprises)

Electric water pumps for Ford street-driver and fully equipped race cars have installation nearly identical to the factory pump and can be completed in 2 to 3 hours. Aftermarket underdrive pulley sets may require a shorter serpentine belt. (Photo Courtesy Meziere Enterprises)

Ford 2011 5OH Meziere Enterprises electric water pump is shown. (Photo Courtesy Meziere Enterprises)

Electric pumps often work better than stock mechanical pumps for street rods. The high-flow electric water pumps will deliver roughly three times the flow of a mechanical pump at idle speed. Street rods rarely have any significant engine load and quite often are cruising at idle speed. Some OEMs, such as Audi, used an auxiliary electric water pump. Brushless DC motor technology now allows OEMs to consider using electric pumps.

Major European manufacturers, such as BMW and Audi, have implemented electric water pumps as standard equipment or as an auxiliary pump on a number of their vehicles. Their research indicates these pumps use 90-percent less energy than mechanical water pumps. They can also provide faster engine warm up and better engine temperature management, eliminate engine heat soak, and improve engine life. BMW has used them on its E82 platform.

There are some controllers available that allow the user to reduce voltage or pulse the voltage at different frequencies to adjust the pump speed to different driving conditions. Newer brushless motor technology has opened up a whole world of possibilities for pump control. The latest Meziere Enterprises remote-mount, brushless pumps come standard with a high- and low-speed switch. They are also capable of communicating with CAN-BUS systems, which can integrate the flow control into the larger control scheme of the vehicle.

One needs to be careful when talking in absolute numbers about pump performance. The standard way to rate a pump was with the free-flow number. That means zero resistance on the exit of the pump and near-zero pressure on the inlet. That can be very tricky to set up in a way that does not bias the test. It also does not tell the whole story about the pump and how it performs when back pressure is presented or inlet restrictions are introduced.

The electric pump cannot compete with the flow of a mechanical pump at high RPM. One major factor is that the electric pump consumes a known quantity of electric power, about 200 watts, no matter what the engine is doing. The mechanical pump consumes as much as 15 to 20 hp at high RPMs. The horsepower is converted to flow and increased block pressure. This block pressure can be a great benefit in towing applications, boosted applications, and other high-horsepower applications where there is continuous wide-open-throttle use.

The additional block pressure effectively raises the boiling temperature of the coolant and prevents hot spots in the cylinder head. Please note that mechanical pumps exceed 110 gpm, but to get that level of performance from an electric pump would not be possible. Meziere Enterprises' best electric pump is rated at 55 gpm free-flow.

Meziere Enterprises has observed about 22 gpm through a standard big-block Chevrolet and 17x24 radiator. Belt- or gear-driven pumps can turn at 4,500 to 6,500 rpm. Brush-type pumps run at about 2,000 rpm, and the new brushless pumps run at about 3,200 rpm. All of the Meziere Enterprises electric water pumps use a scroll-type impeller.

Electric Water Pump Relay Kit

Using a relay when wiring your electric water pump can save you from overloading existing wires and supply the pump with ample power. Meziere Enterprises offers a kit designed for Ford modular installations with wires cut to length. The kit can be used for any of their other electric water pumps.

Meziere Enterprises Belt-Driven Mechanical Water Pumps

Meziere Enterprises also offers mechanical belt- and gear-driven water pumps, which are available for big- and small-block Chevy engines (standard and reverse rotation). This is a highly effective street pump in a full race, 4-inch scroll-type impeller mechanical pump design with impressive operating figures.

These belt- or gear-driven pumps can turn up to 6,500 rpm with a flow rate of 120 gpm consuming about 6.25 hp. They use both low- and high-pressure ports for auxiliary plumbing. These pumps are all-billet construc-

Meziere Enterprises also manufactures mechanical belt- and gear-driven 400-series water pumps for use on GM LS1–LS5 V-8 high-performance engines in production and crate versions, as well as other applications. This pump boasts higher flow with a 4-inch impeller. It is for standard rotation, serpentine-style applications. (Photo Courtesy Meziere Enterprises)

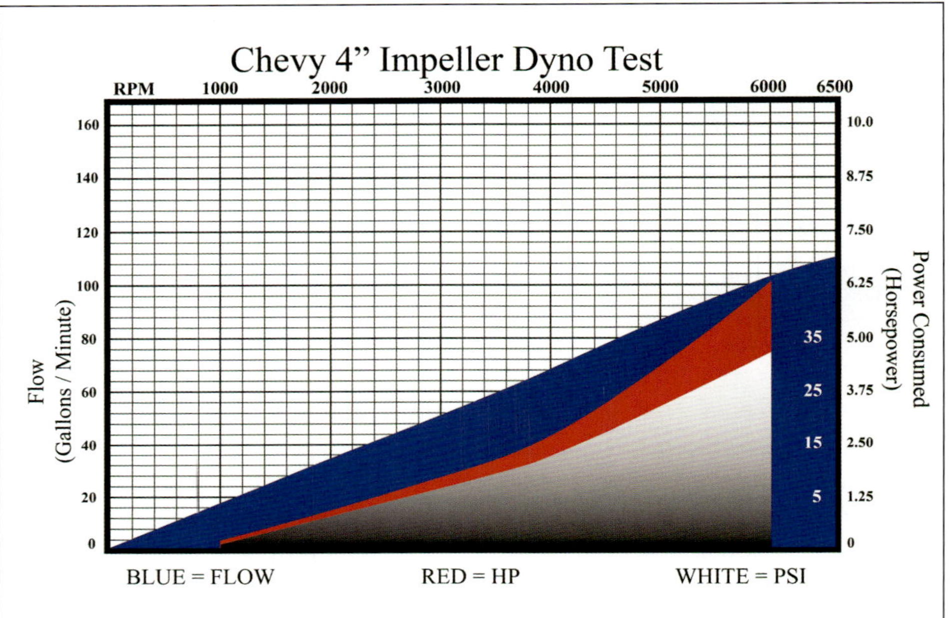

This is a pump performance graph for Meziere Enterprises' mechanical belt- and gear-driven 400-series water pump for use on GM LS1–LS5 V-8 engines in both the production and crate versions. The graph shows that the belt- or gear-driven pumps can turn up to 6,500 rpm with a flow rate of 120 gpm consuming about 6.25 hp. (Chart Courtesy Meziere Enterprises)

The Meziere WP419 mechanical water pump for GM LS1–LS5-series engines is an example of the 400-series belt-driven pumps designed for racing and modified stock. These pumps are all-billet construction and are designed with top-end performance and longevity in mind. Top-end figures match the best racing pumps with an off-idle flow of 5 to 7 gpm. (Photo Courtesy Meziere Enterprises)

tion. The low-speed flow numbers make them popular with street rod builders. The high-RPM performance is capable of cooling any race engine.

Thermostats

The specific function of the coolant thermostat is a flow component in the coolant process. The thermostat is also used to reduce engine warm-up time, maintain engine operating temperature, and heat the inside of the vehicle. A thermostat is about 2 inches wide and regulates coolant flow through the engine. When the engine is cold and the thermostat is closed, no coolant flows through the engine. In some applications, it circulates through a bypass hose or passage. The opening temperature is stamped on the top of the thermostat.

The thermostats consist of a valve that is opened by a thermal expansion device that is generally a wax pellet, and it is closed by the return spring. Valve opening is directly related to coolant temperature around the thermostat. The thermo-

Wax Thermostatic Element

The wax thermostatic element was invented in 1934 by Sergius Vernet (1899–1968). Its principal application is in automotive thermostats used in the engine cooling system. The first applications in the plumbing and heating industries were in Sweden (1970) and in Switzerland (1971).

Research in the 1920s showed that cylinder wear was aggravated by condensation of fuel when it contacted a cool cylinder wall. This contact removed the oil film. The development of the automatic thermostat in the 1930s solved this problem by ensuring fast engine warm-up. Work on cooling high-performance aircraft engines in the 1930s led to the adoption of pressurized cooling systems, which became common on cars postwar.

As the boiling point of water increases with increasing pressure, these pressurized systems could run at a higher temperature without boiling. This increased the working temperature of the engine, thus its efficiency. It also increased the heat capacity of the coolant by volume, allowing smaller cooling systems that required less pump power. The Cadillac Motor Car Division of General Motors was the first OEM to use a thermostat in 1914 on its flathead V-8. ■

stat is usually installed between the cylinder head coolant outlet and the inlet to the radiator, but there are other locations.

Generally, the thermostat was located between the engine hot coolant outlet and the radiator inlet, which is the pressure side. However, it can be located on either the suction or pressure side of the water pump. For example, the LT1 engine used in the Corvette from 1992 to 1996 had the thermostat located on top of the water pump. It was actually part of the water pump. And the later LS1-LS5-series GM engines had the thermostat located on the suction side of the cooling system as part of the water pump inlet coming from the bottom of the radiator and being sucked into the engine.

There are also high-performance applications with a remote-located thermostat directly in the coolant return hose to the radiator. There are some applications with a vapor line thermostat outlet to the water pump to eliminate any cavitation as well.

Most early domestic engines had the thermostat on the pressure side at the top of the intake manifold. Some engineers believed that the thermostat should not be placed on the suction side of the water pump because the coolant will be drawn through a restriction. The general thinking is that the placement of the thermostat does not affect the engine operating

This cutaway shows the location of the thermostat in the outlet coming from the cylinder heads. The thermostat is a temperature-sensitive valve that regulates the flow of coolant between the engine and the radiator. Since the engine creates heat, it is designed to operate at high temperatures. When the engine is cold, it must be warmed quickly to its ideal operating temperature. To speed the warming up, the thermostat stays closed when the coolant is cold. (Photo Courtesy Jim Halderman)

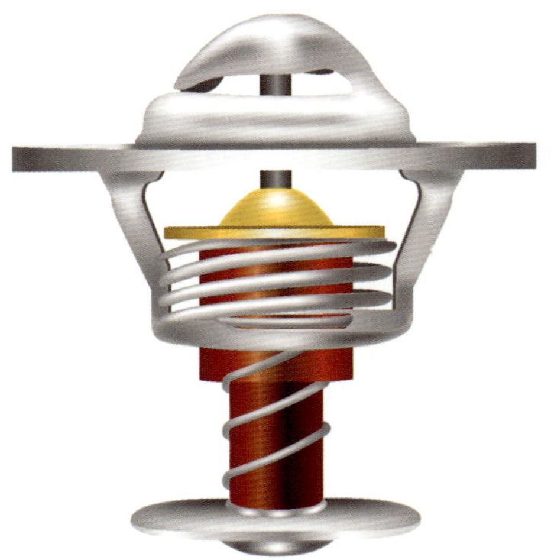 

Engines that are sensitive to thermal shock caused by surges of coolant require a tighter control of temperature. They may use a constant inlet temperature system. In this arrangement, the inlet cooling to the engine is controlled by dual-valve thermostat (left) that mixes a recirculating sensing flow with the radiator cooling flow. These employ a single capsule but have two valve discs. They are compact and simple but an effective control function is achieved. On the right is a standard thermostat design.

In the LT1 engine used in the Corvette from 1992 to 1996, the thermostat is located on top of the water pump and is actually part of the water pump. (Photo Courtesy Jim Halderman)

Derale Performance LS1-LS5-series GM engines have the thermostat location as part of the water pump. It is located on the suction side of the cooling system on the inlet coming from the bottom of the radiator and being sucked into the engine. (Photo Courtesy Derale Performance)

temperature. For example, a 180°F opening thermostat placed on the pressure side of the cooling system would need to be calibrated to 260°F if placed on the suction side in order to create identical engine operating temperatures.

Thermostat Operation

The standard thermostat has a small cylinder that is filled with a wax that begins to expand at whatever the calibrated opening temperature is set to. A rod connected to the valve presses into this wax. When the wax gets hot, it expands, pushing the rod out of the cylinder and opening the valve to allow coolant flow to the radiator.

The thermostat control takes place at the opening temperature and will control coolant flow until it is fully open. Once engine temperatures reach the normal operating range, the cooling system maintains the appropriate temperature range to prevent excessive heat and pressure buildup. The coolant temperature rises and falls as it circulates through the system.

As the coolant temperature rises, the thermostat opens and the coolant is directed into the radiator. When the coolant temperature falls, the thermostat closes slightly and decreases the flow of coolant through the radiator. The thermostat's position varies during normal operation, but it is seldom open completely. Operating conditions that can open the thermostat completely include long uphill climbs, idling in heavy traffic, or extremely hot days. Thermostats are available with different opening temperatures. Common settings are in the range of 180°F to 195°F (82°C to 90°C). Thermostats should be fully open about

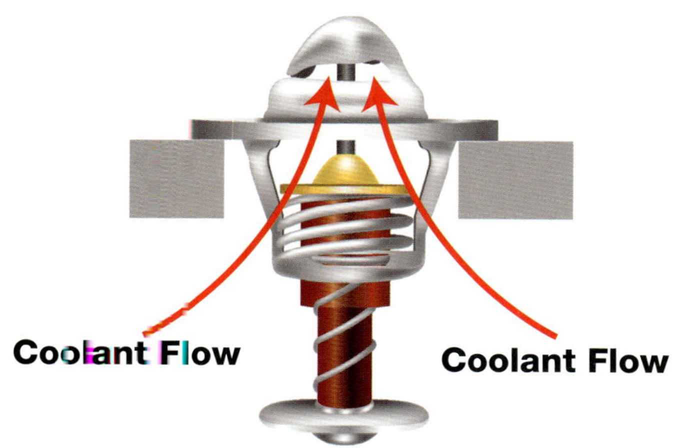

Wax pellet-type automotive thermostats have a small cylinder that is filled with wax that begins to expand when the calibrated opening temperature is reached. A rod connected to the valve presses into this wax. When the wax gets hot, it expands and pushes the rod out of the cylinder, compressing the return spring and opening the valve to allow coolant to flow into the radiator.

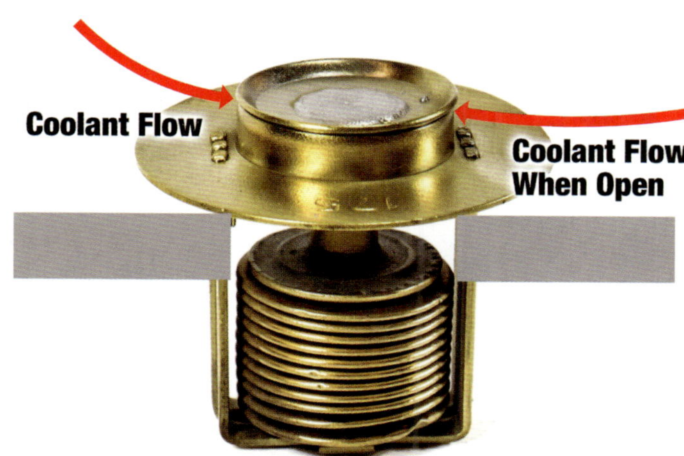

A bellows-type engine thermostat used a bellows to open and close the valve that allows coolant to flow to the radiator. A volatile liquid, such as Benzene, was inside the bellows, and it expanded when heated to move a rod upward. The rod was attached to the valve that opens a passage to the radiator.

20°F (−1°C) higher than their rated opening temperature.

The opening temperature is the key element when selecting a replacement thermostat. The thermostat must open at the specified temperature for the particular system being serviced. When the opening temperature is too low, the thermostat opens too soon and the warm coolant is circulated out of the engine prematurely. This can cause reduced engine performance and sludge formation in the crankcase. The engine management system will see the cold engine temperature as a fault, causing the malfunction indicator lamp (MIL) or check engine light to come on and setting a diagnostic trouble code (DTC). This will also prevent normal operation of the heater and defroster.

If the opening temperature is too high, the thermostat does not open fast enough. In that case, the hot coolant remains in the hot engine instead of being circulated to the radiator to be cooled. An engine that is operated at too high a temperature can cause various problems including engine ping, detonation, increased oil consumption, or reduced engine life.

Thermostats can be damaged by the corrosion and scale that build up in the cooling system. This scale can attach to the stem of the thermostat, causing sticky valve movement. A thermostat that sticks open or shut can cause the same problems as a thermostat with an incorrect opening temperature.

Most engines are equipped with a thermostat bypass circuit. The system permits coolant to flow through the water pump and back into the engine when the thermostat is closed. The bypass itself may be an external hose or a passage within the thermostat housing or water pump body.

Early thermostats had a bellows design in which the coolant temperature expanded a bellows to open the thermostat. A drawback to the bellows-design thermostat was that it was sensitive to pressure changes, which meant it could sometimes be forced shut again by pressure. This led to overheating with a banging noise. The wax pellet type has a negligible change in its external volume, so it is insensitive to pressure changes. Many 1950s-era cars were originally built with bellows thermostats and later serviced with replacement wax capsule thermostats without requiring any adaption.

Electronic Thermostat

An electronic thermostat allows the engine temperature to be more closely controlled during cruise, city driving, or heavy-load conditions to optimize fuel economy and reduce exhaust emissions. The engine control module (ECM) operates the electronic thermostat based on input sensor information from the engine

---

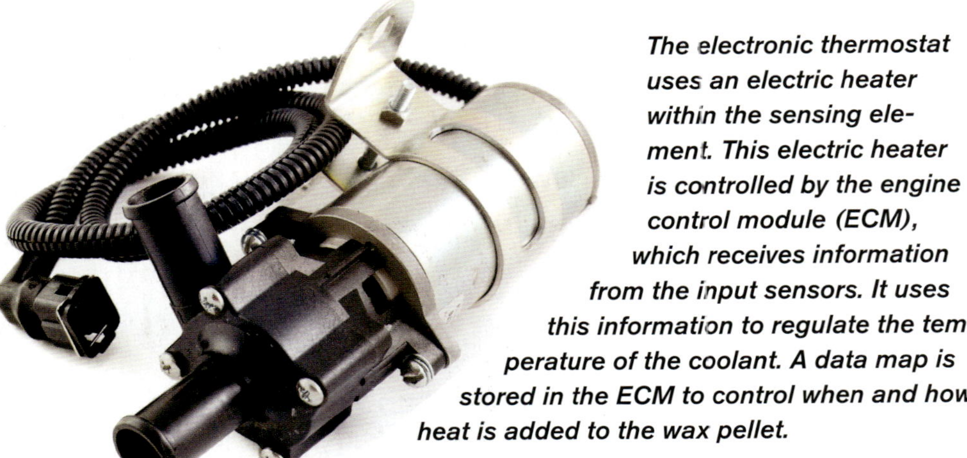

The electronic thermostat uses an electric heater within the sensing element. This electric heater is controlled by the engine control module (ECM), which receives information from the input sensors. It uses this information to regulate the temperature of the coolant. A data map is stored in the ECM to control when and how heat is added to the wax pellet.

coolant temperature (ECT) sensor or the air fuel ratio sensor (old oxygen sensor), crankshaft position (CKP) sensor (for RPM), and mass airflow (MAF) sensor for engine load.

The electrically assisted thermostat provides broader operation than traditional thermostats. In addition to the mechanical function of the wax pellet, it uses an electric heater within the sensing element. This heater is controlled by the ECM, which receives information from the input sensors. It uses this information to regulate the temperature of

the coolant. A data map is stored in the ECM to control when and how heat is added to the wax pellet.

Thermostat Bypass Circuit

Almost all engines use a bypass circuit to speed up the engine warm-up process. Some applications use a bypass hose that circulates coolant from the hot engine back into the water pump outlet passage. It allows coolant from the intake manifold before the thermostat opens to flow back into the system.

Other applications, such as the Corvette LT1, use a bypass passage at

the bottom of the thermostat housing where the valve allows coolant from the intake manifold and heads to enter the water pump inlet and be pumped back into the engine. When it is closed, this passages allows coolant to flow or circulate through the open thermostat and back into the radiator for cooling.

The GM Duramax 6600 6.6L diesel uses two thermostats to provide improved temperature control. The rear or primary thermostat is a non-blocking, two-stage design. The first stage of the rear thermostat begins to open at 180°F (82°C), and a small passage is open for coolant to flow to the water outlet and begin to circulate to the radiator. The secondary or second stage begins to open at 185°F (85°C), and more coolant is allowed to pass through to the radiator. The rear or primary thermostat is fully open at 203°F (95°C).

The front or secondary thermostat has a single-stage design with a bypass flapper valve. It begins to open at 185°F (85°C). As it begins to open, it also begins to close the

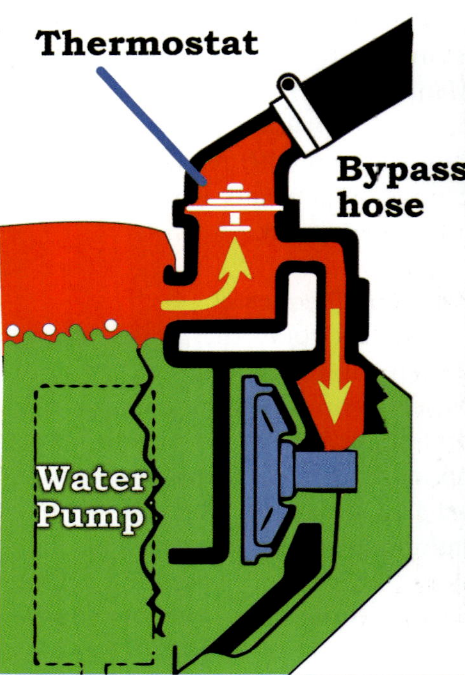

The graphic (left) illustrates the thermostat bypass hose that routes past the thermostat for faster engine warm-up. On the right is a photo of a Corvette LT1 bypass passage located below the thermostat, also for a faster warm-up.

The GM Duramax 6600 6.6L diesel uses two thermostats to provide improved temperature control. The rear or primary thermostat is a non-blocking, two-stage design. (Photo Courtesy Jim Halderman)

bypass flapper valve and the bypass passage. The front thermostat is fully open at 212°F (100°C).

Air in the Cooling System

A cooling system leak will not directly cause an engine to overheat the direct cause is air entering the cooling system. When the system is blocked and the coolant cannot circulate through the radiator to disperse heat, air is trapped and forms air pockets or hot spots. A hot spot is an area of trapped pockets of air or bubbles inside cooling system passages that prevent coolant from carrying heat away from those areas, causing the engine to overheat. These spots can cause overheating cracks or warp any part of the engine.

Any compressed substance left in the coolant, such as air (which is the biggest enemy of a cooling system), can cause overheating and seriously damage both the engine and the water pump. Air-accumulating convention can form an insulating sublayer at the outside of the combustion surface. This insulating layer prevents heat transfer from the hot surface to the flowing coolant, interfering with cooling. Boiling is harmful because the gases form steam pockets in the water jackets. Steam in the cooling system absorbs much less heat than liquid, so localized hot spots may form and can warp or crack castings.

Nucleate Boiling

Nucleate boiling takes place when the overall temperature of the coolant stays below its boiling point with only a small nuclei of coolant actually touching the metal, which flashes into a gas. For water, nucleate boiling takes place when the surface temperature is greater than the saturation temperature (between 18°F and 54°F). The heat transfer from surface to liquid is greater than that in film boiling. Some of the coolant is actually boiling all of the time. Should the process stop, the engine would rapidly overheat and seize.

Nucleate boiling is one of the keys to low engine temperature. Ordinary ethylene glycol and water coolant boils at around 263°F. However, in the cylinder head, the temperature of the combustion chamber or exhaust port walls is greater than 500°F. As droplets of coolant flow through the water jackets against the other sides of these hot surfaces, they boil into a gas and are instantly washed away, to be replaced by coolant. The superheated bubbles are swept into the lower-temperature coolant flowing through the middle of the passage, where they cool off and condense back to liquid. Boiling the coolant into a gas removes heat from the metal surface much more efficiently than merely warming the coolant.

Nucleate boiling requires a strong, turbulent coolant flow to perform properly. If the coolant were to stop moving, steam pockets would form quickly and result in overheating, detonation, boilover, and warped or cracked castings. Many engine blocks and cylinder heads have special passages called steam holes that permit bubbles from boiling coolant to escape when the engine is operated at low speeds or turned off. When boiling happens, an additional superheated vapor film is formed. The surface temperature continues to rise into a hotspot. Evans waterless coolant helps in eliminating these boiling problems.

Gaseous coolant entering the suction side of the water pump causes low pressure and rapid loss in the pump capacity by increasing the minimum net positive suction head (NPSH). NPSH designates the lowest inlet pressure required by the water pump at a given flow to avoid cavitation. The gaseous coolant lowers the boiling point of the coolant, which lowers the normal vapor pressure boiling point in the coolant, which will begin to boil.

It's extremely important that you get all the air out when filling the cooling system. You must run the engine long enough for that to take place. You also need to bleed any hotspots that have bleeder valves. You must open the bleeder valve to get all the air out of cooling system with the engine running.

Bleeding

The thermostat uses a valve called the jiggle valve, or a pin at the top, which vibrates as the coolant passes through it and allows air to bleed out of the system. This takes place when the thermostat valve is closed before the engine reaches operating temperature. You could call it a self-bleeding system. However, air located in high spots cannot escape through the thermostat bleed. The high spot in the cooling system must be provided with a second bleed valve (similar to brake systems) to bleed out the air.

Thermostat Selection

Late-model production cars use a calibrated thermostat between 195 and 205°F, while the performance aftermarket offers them in the 160 to 280°F ranges. These ratings specify the temperature that the thermostat will open and start the flow. It takes another 20°F for it to be fully open.

The placement of the thermostat, whether on the pressure or suction side of the coolant flow, will affect the rating. A rule of thumb is a thermostat on the suction side will have an opening temperature 10 to 20 degrees lower than the thermostat on the pressure side. The result will be the coolant entering the engine will be about the same temperature in either case.

Most engines are thermally more efficient at a coolant temperature of 195 to 200°F. You should choose a thermostat near to what the OEM originally used in the engine, or the best bet is a 195°F opening temperature thermostat.

Thermostat Location

The thermostat is a key component in the cooling system because it accelerates engine warm-up

The engine thermostat is the controlling device for engine temperature and the gateway for coolant to flow into the radiator. At the top of the thermostat is a jiggle valve (or sometimes just an orifice), and when the thermostat is closed, air can be bled out of the cooling system. (Photo Courtesy Jim Halderman)

Here is an example of a remote engine thermostat housing as a billet aluminum component. This Meziere unit has a 1.5-inch inlet and outlet, and it offers a very high flow rate when combined with a high-flow thermostat. (Photo Courtesy Meziere Enterprises)

This high-flow Meziere thermostat has a large air bleed on the flat area of the unit, which is perfect for performance applications when combined with the high-flow remote thermostat housing. (Photo Courtesy Meziere Enterprises)

Many OEMs recommend that the bleeder valve be opened whenever refilling the cooling system. For example, Fiat-Chrysler recommends that a clear plastic hose be attached to the bleeder valve and directed into a suitable container to keep from spilling coolant onto the ground and on the engine. This also allows the technician to observe the flow of coolant for any remaining air bubbles. (Photo Courtesy Jim Halderman)

Meziere makes a series of larger high-flow remote thermostat housings with outlets to the radiator and thermostats that complement its line of water pumps. (Photo Courtesy Meziere Enterprises)

A flexible universal upper radiator hose is installed on this modified vehicle. (Photo Courtesy Champion Cooling Systems)

Here is an example of a molded upper radiator hose installed on a modified vehicle. (Photo Courtesy Bob Wilson)

and improves your performance, but it can be a big flow restriction even when fully open. A remote bypass-style thermostat can be used for several reasons. The primary reason is for a higher flow rate.

Meziere makes a series of larger high-flow remote thermostat housings and thermostats. You can also use a larger outlet to the radiator for better coolant flow to reduce any chance of overheating. If your system will not permit the use of a remote or high-flow thermostat, you can drill four 1/8-inch holes around the thermostat perimeter to increase coolant flow.

Coolant Hoses

Radiator hoses connect the cooling system passages to the radiator tanks and heater core. The hoses must be flexible and absorb the motion between the vibrating engine and the stationary radiator. The hoses are typically made of synthetic rubber, and the bottom or suction hose from the water pump is often reinforced with a wire coil to prevent implosion from the suction being squeezed by atmospheric pressure.

Coolant hoses are designed to fit the specific vehicle application. In the past, you could only get a factory hose from the OEM such as General Motors, Ford, or Chrysler. Aftermarket hoses were generally flex hoses that would bend to fit the location, and sometimes they did not work too well. However, in today's market you can buy factory OEM exact duplicate molded hoses to fit your factory application. You can purchase these hoses from rockauto.com, 1A1auto.com, or even Amazon. Yet, creating a custom fitting hose and the AN design for modified stock and race applications may be your best bet.

When replacing cooling system hoses, don't select the cheapest hoses. Spend the extra money on the best components. Goodyear Super Hi-Miler or Gates performance cooling system hoses last longer than your average off-the-shelf hose, especially when paired with high-quality worm gear clamps.

The hose ends fit over necks or fittings on the engine and radiator and are held in place with hose clamps. Hose clamps can be held in place by spring tension or by a screw.

Screw- and gear-style hose clamps offer one of the best designs for easy installation, removal, and tension.

Shown here is a typical AN flare end fitting for flexible upper radiator hoses.

This modified vehicle is using AN fitting coolant hoses connected from the engine coolant outlet to the downflow radiator inlet. *(Photo Courtesy Champion Cooling Systems)*

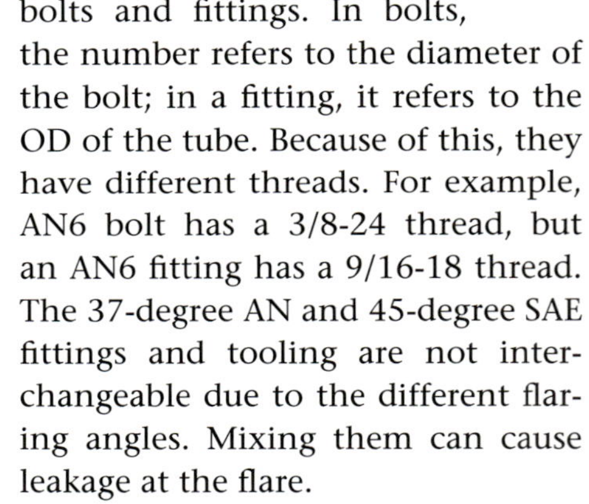

The "AN" Thread

The AN thread is a type of fitting used to connect flexible hoses and rigid metal tubing that carry fluid such as engine coolant. It is a US military–derived specification that dates back to World War II and comes from a joint venture by the army and navy, hence the term AN.

AN sizes range from -2 (dash two) to -32 in irregular steps, with each step equating to the outside diameter (OD) of the tubing in 1/16-inch increments. So, an -8 AN size would be equal to 1/2-inch OD tube (8 × 1/16 = 1/2). Yet, this system does not specify the inside diameter (ID) of the tubing because the tube wall can vary in thickness. Each AN size also uses its own standard thread size.

AN fittings are a flare fitting using 37-degree flared tubing to form a metal-metal seal. They are similar to other 37-degree flared fittings, such as JIC, which is the industrial variant. The two are interchangeable in theory, though this is typically not recommended due to the exacting specifications and demands of the aerospace industry. The differences between them relate to thread class and shape (how tight a fit the threads are) and the metals used.

The AN threads are different for bolts and fittings. In bolts, the number refers to the diameter of the bolt; in a fitting, it refers to the OD of the tube. Because of this, they have different threads. For example, AN6 bolt has a 3/8-24 thread, but an AN6 fitting has a 9/16-18 thread. The 37-degree AN and 45-degree SAE fittings and tooling are not interchangeable due to the different flaring angles. Mixing them can cause leakage at the flare.

AN Sizes														
	-2	-3	-4	-5	-6	-8	-10	-12	-16	-20	-24	-28	-32
Tube OD	1/8 inch	3/16 inch	1/4 inch	5/16 inch	3/8 inch	1/2 inch	5/8 inch	3/4 inch	1 inch	1¼ inches	1½ inches	1¾ inches	2 inches
Nominal Hose ID	1/8 inch	3/16 inch	1/4 inch	5/16 inch	3/8 inch	1/2 inch	5/8 inch	3/4 inch	1 inch	1¼ inches	1½ inches	1¾ inches	2 inches
Actual Hose ID (approximate)	1/16 inch	1/8 inch	3/16 inch	1/4 inch	5/16 inch	7/16 inch	1/2 inch	5/8 inch	7/8 inch	1⅛ inches	1⅜ inches	1⅝ inches	1¾ inches
SAE Thread Size	5/16-24	3/8-24	7/16-20	1/2-20	9/16-18	3/4-16	7/8-14	1¹⁄₁₆-12	1⁵⁄₁₆-12	1⅝-12	1⅞-12	2¼-12	2½-12
Pipe Thread Size (NPT)	–	1/8-27	1/4-18	–	3/8-18	1/2-14	–	3/4-14	–	–	–	–	–

RADIATOR AIRFLOW

Airflow through and under the radiator is one of the three basic parameters that determine cooling efficiency. The demand for more-powerful engines in smaller hood spaces has created a problem of insufficient rates of heat dissipation through the radiators. Upward of 33 percent of the energy generated by the engine through combustion is lost in heat.

Insufficient heat dissipation can result in the engine overheating, which leads to the breakdown of lubricating oil, weakening of metal engine parts, and significant wear between engine parts. To minimize heat generation, there must be sufficient airflow through the radiator.

Airflow

The three basic areas that determine engine cooling efficiency are: radiator surface area, coolant speed through the system, and the amount of airflow through the radiator. These factors determine the efficiency of a cooling system as expressed in British Thermal Units (BTU) of heat rejection per minute. The biggest limitation in terms of radiator surface area is the original vehicle itself.

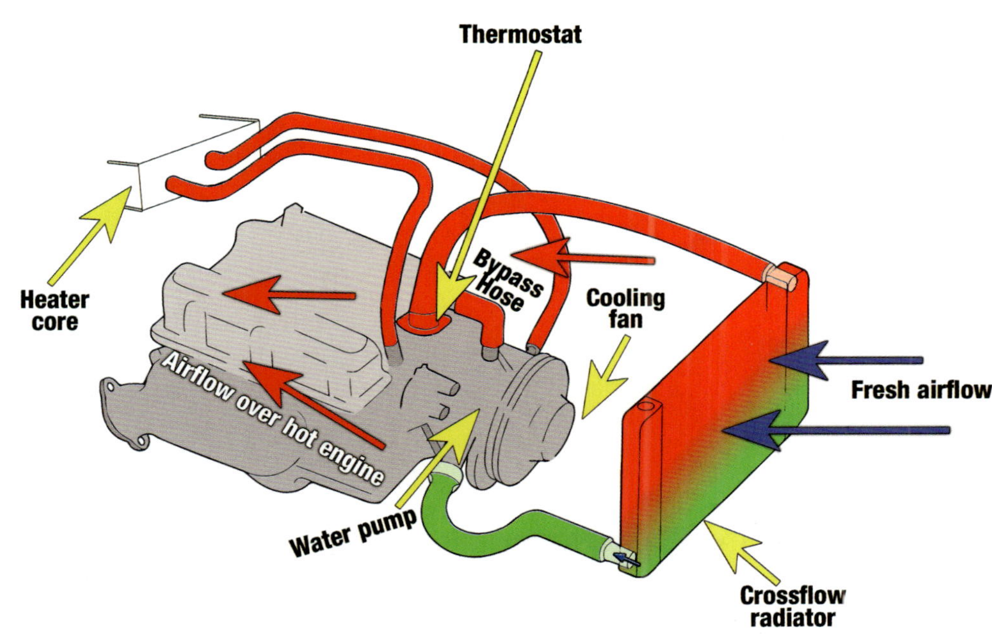

The cooling system airflow begins as air pressure flows through and under the vehicle (also over for aerodynamics). The airflow through the radiator is a determining factor in cooling system design and preventing overheating. This illustration shows that there is no radiator shroud, which is a critical component of cooling system airflow.

The core support will usually determine the size of the radiator. When the OEMs develop a cooling system, they must create a balance where the radiator is big enough to control the engine's heat yet still be able to fit in the vehicle and be inexpensive.

Airflow is the movement of air from one area to another. The primary cause of airflow is the existence of pressure inclines. Air behaves in a fluid manner, meaning particles naturally flow from areas of higher pressure to lower pressure. Atmospheric pressure (14.7 psi) is directly related

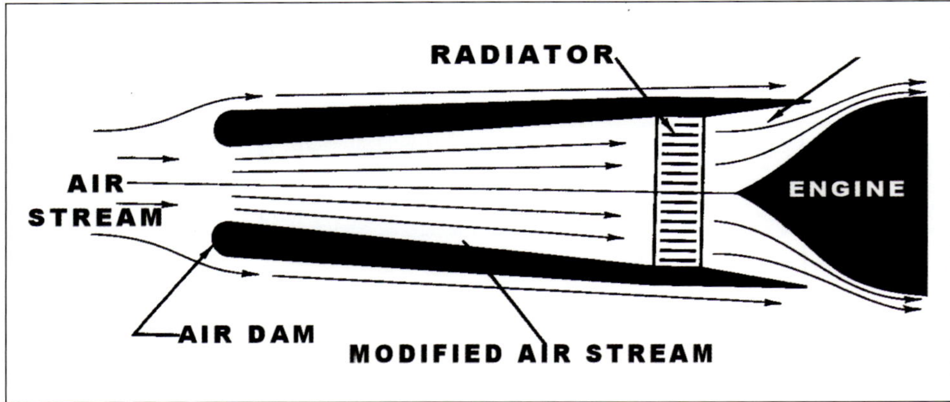

Airflow pattern through the radiator and adequate cooling airflow through the engine radiator are essential elements of vehicle thermal management.

This combination air dam and radiator air intake is on a modified Chevrolet Silverado. This air dam keeps the front from lifting off the ground and provides cool airflow through the radiator from the air inlets below the grille. (Photo Courtesy Jim Halderman)

related to front-end appearance, as well as the trade-offs with aerodynamic performance.

The use of more-powerful engines in modified vehicles with higher heat rejection requirements and ever tighter underhood packaging pose significant challenges in the design of efficient cooling airflow systems. As the vehicle modifier, you must also make sure the components, such as the engine and the transmission, will operate within the specified temperature ranges. Solid prediction of the cooling of thermal components and systems early in the development process is important.

Airflow Types

Like any fluid, air may exhibit both laminar and turbulent flow patterns. Laminar airflow is when each air particle follows in a smooth path and the paths never interfere with each other. The air velocity of the fluid is constant at any point in the airflow. Turbulent flow is irregular flow that is characterized by tiny whirlpool regions. The velocity of this fluid is definitely not constant at every point. Turbulent flow occurs when there is an irregularity (such as a disruption in the surface across which the air is flowing) that alters the direction of movement.

Laminar flow occurs when air can flow smoothly and exhibits a parabolic velocity profile. A parabolic (resembling the geometric figure parabola) velocity profile is used to model or measure laminar (or smooth) airflow. The resistance to flow can be characterized in terms of the viscosity of the air if the flow is smooth. In the case of a moving plate in the air, it is found that there is a layer, or lamina, that moves with the plate, and a layer that is essen-

to altitude, temperature, and composition. In engineering, airflow is a measurement of the amount of air per unit of time that flows through a particular device. The flow of air can be induced through mechanical means (such as by operating an electric or manual fan) or can take place passively (as a function of pressure differentials present in the engine compartment).

Adequate cooling airflow through the engine radiator is an essential element of vehicle thermal management. The design of an automobile's cooling airflow management system should also account for the constraints and design issues

tially stationary if it is next to a stationary plate.

Reynolds Number

The Reynolds number is an engineering term for a quantity with no dimensions that is used in fluid mechanics to help predict airflow patterns. At low Reynolds numbers, flows tend to be dominated by laminar flow; at high Reynolds numbers, turbulence results from differences in the air speed and direction that may sometimes intersect or even move counter to the overall direction of the airflow.

The Reynolds numbers are used to measure a range or situations, from liquid flow in a pipe to the passage of air through an engine cooling radiator. It is used to predict the transition from laminar to turbulent flow. The predictions of the beginning of turbulence and the ability to calculate scaling effects can be used to help predict airflow behavior.

The speed at which the air flows past an object varies with distance from the object's surface. The positions of the radiator, fan, and shroud are important to achieve the maximum cooling. The region surrounding an object where the air speed approaches zero is known as the boundary layer. It is here that surface friction most affects flow; irregularities in surfaces may affect the boundary layer and disrupt airflow.

Airflow Measurement

An anemometer measures airflow and is sometimes called an airflow meter. This tool may use ultrasound or resistive wire to measure the energy transfer between the measurement device and the passing particles. A hot-wire anemometer, for example, registers decreases in wire temperature, which can be translated

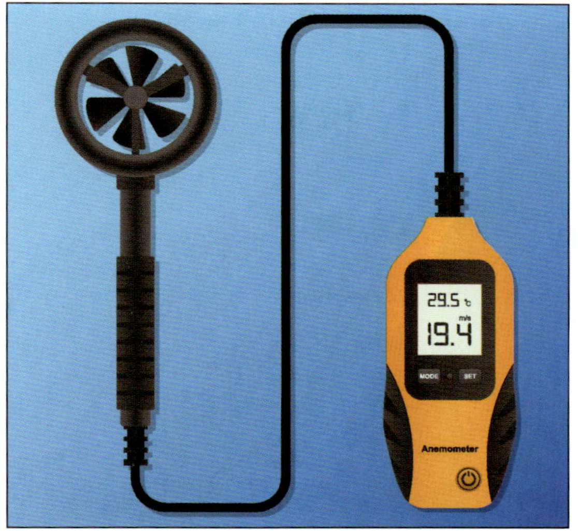

Anemometers are also used to measure wind speed and indoor airflow. There are a variety of types, including straight probe anemometers, which are designed to measure air velocity, differential pressure, temperature, and humidity. Rotating vane anemometers are used for measuring air velocity and volumetric flow. (Photo Courtesy Jim Halderman)

into airflow velocity by analyzing the rate of change. I am not suggesting that any modified or race builder would use one. It is mentioned so you would know the science behind measuring airflow.

Airflow to Remove Heat

Vehicles used for racing or modified cars can develop more heat than a stock street vehicle. Although, today OEMs build vehicles that have race car features, such as the Chevrolet Corvette, Dodge Hemi, and Ford Mustang. This heat generation occurs due to many different factors, including high RPM operation for longer periods than stock vehicles and radical camshaft profiles. For example, NASCAR engines produce about 750 hp and use extremely radical cam

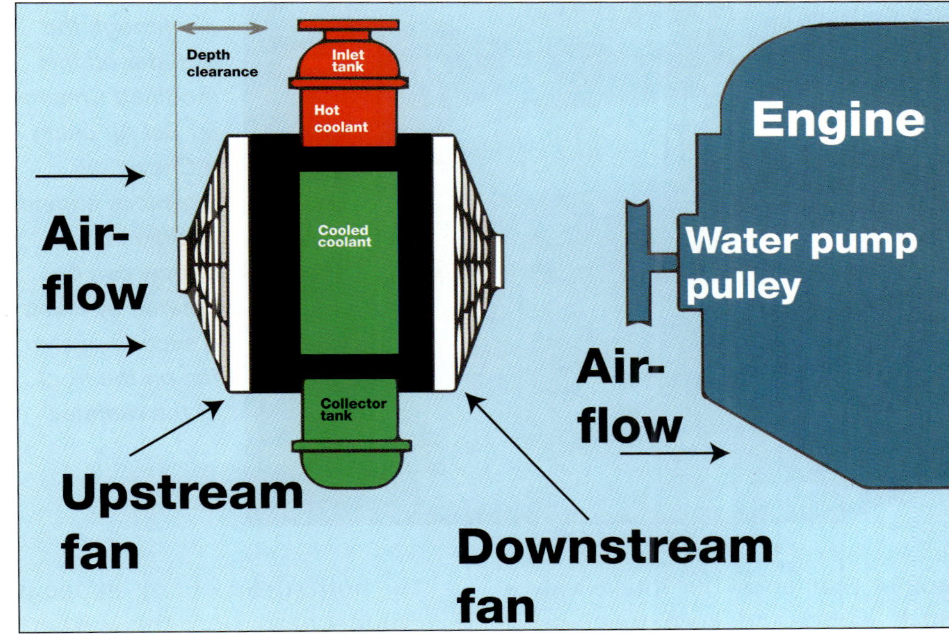

This illustration shows both a downstream fan and an upstream fan, where the downstream fan is sucking the air through the radiator in the natural airflow stream and the upstream fan is pushing the air through the radiator.

This is a common downstream fan, which sucks air through the radiator in the natural airflow stream on this modified Chevrolet Bel Air using a Chevy 396 big-block engine. An electric fan placed on the engine side of the radiator (as a puller) is always more efficient than a pusher fan.

An upstream fan pushes the air through the radiator on this modified Chevrolet Bel Air using a Chevy 396 big-block engine. Additional airflow can be created by using a second pusher fan on the front of the radiator.

the chances of a head gasket failure. As a cylinder head heats more than the cylinder block, it will expand farther and slide across the head gasket more. When the head does this many times, the gasket is more likely to fail. This is called micromotion.

Accurate prediction of cooling airflow is needed to consistently predict the performance of the engine under actual vehicle operating conditions. The flow patterns and the air temperature distribution in the underhood are highly complex. They are constantly changing because of the intricate interaction of the ram air and the cooling fan flows along with variations in air density and temperature as a result of heat from the engine. In addition, they are very sensitive to vehicle operating conditions and fine geometric details. You need a straight path for the air to flow through the radiator.

Other factors influence the temperature of the engine, including radiator size and the type of radiator fan. The size of the radiator and its cooling capacity is chosen so that it can keep the engine at the operating temperature under the most extreme conditions a performance vehicle is likely to meet. Airflow speed through a radiator is a major influence on the heat it loses. Vehicle speed affects this in proportion to the engine effort, thus giving self-regulatory feedback. Where an additional cooling fan is driven by the engine, this also tracks engine speed similarly.

Engine-driven fans are often regulated by a viscous-drive clutch from the drive belt, which slips and reduces the fan speed at low temperatures. The viscous clutch is a fluid coupling that uses silicone fluid as the connecting fluid. Further regulation of cooling rate is provided by

profiles that open the intake valves much earlier and keep them open longer than street cars. This allows more air to be packed into the cylinders, especially at high speeds.

The hottest part of any engine is the cylinder head, since this is where the combustion chamber is located, running over 5,000°F. High cylinder head temperatures tend to increase

This modified Chevrolet Corvette has a viscous clutch mechanical cooling fan and factory shroud.

This modified Oldsmobile stock vehicle uses dual electric cooling fans that are thermostatically controlled by a sensor in the cylinder head.

either variable speed or cycling radiator fans.

Electric fans are controlled by a thermostatic switch, the engine management computer, or engine control module (ECM). Electric fans also have the advantage of giving good airflow and cooling at low engine RPM or when stationary, such as in slow-moving traffic. If the vehicle was originally equipped with an air scoop or air dam, never remove it. More than likely, it forces airflow across the radiator.

Before the development of viscous-drive and electric fans, engines were fitted with simple fixed fans that drew air through the radiator at all times. Vehicles with this design required the installation of a large radiator to deal with a high-horsepower engine. They would often run cool in cold weather under light loads, even with the presence of a thermostat, because the large radiator and fixed fan caused a rapid and significant drop in coolant temperature as soon as the thermostat opened.

Radiator Shrouds

In just about all high-horsepower performance-based cases, cooling fans should be shrouded for proper vectoring of air velocity through the radiator. I recommend paying close attention to what the factory does in any application relative to your modified or race application.

The primary purpose of a fan shroud is to direct the airflow over the radiator. You can experience heat issues when you do not use one on a high-horsepower engine. The shroud directs the incoming cool air directly at the radiator and helps reduce previously heated air from circulating back through the radiator.

Airflow through the radiator and condenser is vital for engine cooling. The shroud and air dams are responsible for 10 to 30 percent of the system's performance. Air shrouds and dams create low pressure behind the radiator and high pressure in front, drawing cool air through the condenser

A radial fan without a shroud can be found on very early automobiles (left). The air is not channeled and it goes in a radial direction away from the fan blades. This can also take place if the shroud is loose or too large. A fan shroud installed on a modified car (right) has the airflow directed through the shroud and the radiator and over the engine.

This early vehicle uses a nonviscous-clutch direct drive cooling fan that operates at engine speed with no variability or control. There is no shroud to channel the airflow.

A modified builder went to extraordinary lengths to ensure that all the air entering the grille would travel through the radiator. Building small-block off-plates between the radiator and the core support to direct all the air through the radiator will accomplish the same effect.

and radiator fins. The lack of a fan shroud could reduce cooling by up to 30 percent. The shroud can only help improve the airflow through your radiator, so it is best to use one.

Under Vehicle Air Dams

Air dams improve airflow at highway speeds because they direct airflow and create a low pressure area that directs the air through the radiator at high speeds. Airflow created by vehicle movement is considered ram air because it is forced through the radiator through vehicle movement.

An ideal ram air ducting radiator setup might look like this:

1. The air entrance has well-rounded leading edges to reduce the possibility of separation and the dock where it is at has a very modest yawl or pitch angle.

2. You will need a long diffuser with a small diffuser angle of about 5 degrees. This theoretically enables the flow of air to slow down and build up pressure

just before the radiator without separation of the airflow. The design could bring the velocity down to a level compatible with the radiator core heat transfer characteristics.

3. The radiator will fit tightly in the dock and all of the incoming air must go through the radiator.

4. Lastly, the contraction section to the rear of the radiator will speed up the air to near free stream velocity so it can be ejected with minimum loss. This would be the air that will flow over the top of the engine.

This will most likely never be able to take place except in the most ideal OEM situations. There is seldom enough space to include a long diffuser. Long diffusers have boundary layer buildup that can lead to separation except under the most-perfect conditions. This will decrease the size of the column of air as it approaches the radiator. My experience is that short diffusers are the best approach, as is the case in most of the OEM vehicles.

On the nozzle side, packaging can be limited due to space. In many cases, all that can be done is to avoid blocking the flow of air and provide as much contraction as possible in the air exit (air coming through the shroud where the fan lives).

The rational design of an engine cooling system is a very complex task. It has been accomplished by the automotive industry only in the last few decades. There are three constraints that are closely related to the cooling performance of a radiator:

- the core pressure drop coefficient (this is not important to us because we are not going to design the radiator; that has

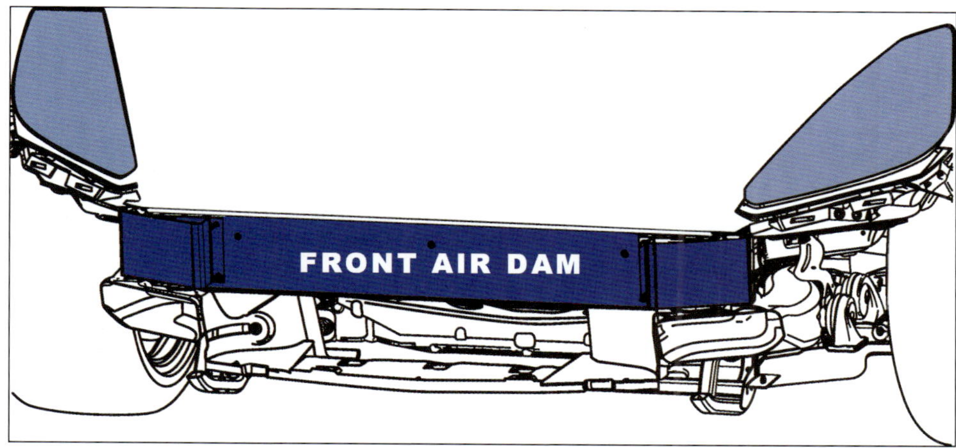

This production vehicle uses a ram air ducting radiator with a lower air dam that closes off the bottom of the vehicle. All of the airflow through the vehicle's grille goes though the radiator.

already been done for us by radiator manufacturers)
- the velocity ratio of the airstream
- the drag coefficient of the radiator

The third generation (1982–1992) Chevrolet Camaro or *F* car came with an air dam placed directly under the radiator. On older, high-mileage cars, this might have been damaged or removed. These air dams were essential to creating a low-pressure area behind the radiator to move air through the radiator.

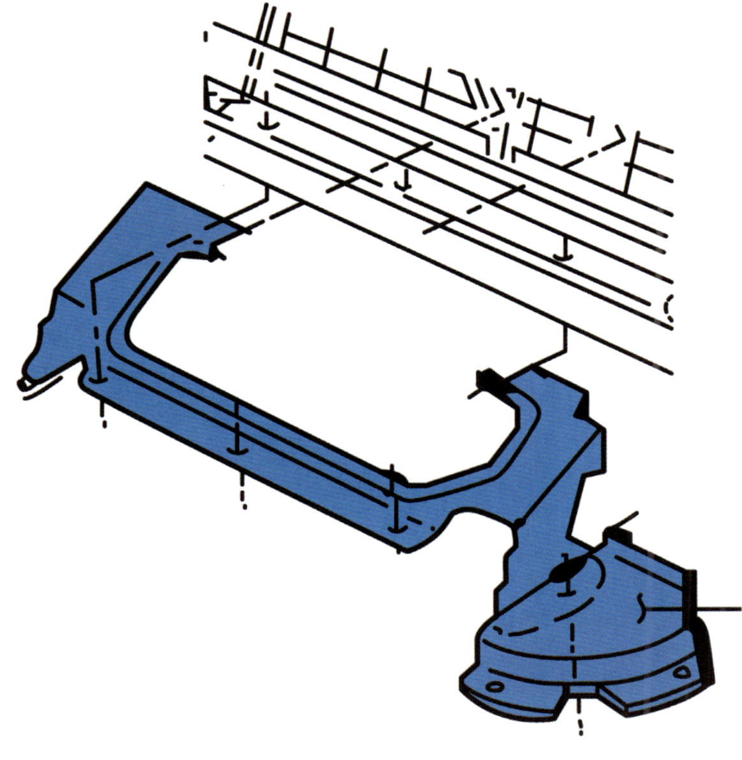

A 1996 Chevrolet Camaro front air dam was placed directly under the radiator. It was essential to create a low-pressure area behind the radiator in order to move air through the radiator.

COOLING FANS

Engine cooling fans are used to draw air through the radiator while the vehicle is at idle or moving at a slow speed. When the vehicle is moving at sufficient speed, air is being forced through the radiator at greater cubic feet per minute (CFM) than the cooling fan is capable of, making the fan somewhat excessive. There are four individual types of engine cooling fans: fixed, flex, viscous clutch, and electric. Each has its own characteristics, as well as drawbacks.

General Cooling Fan Information

Automotive radiators are generally placed at the front of the vehicle, and there may be enough ram air forced through the radiator at cruising speeds so that the fan will not be powered under cruise conditions. Some OEMs design their vehicles so that the amount of ram air is quite large, which allows the fan to be relatively small. Other OEMs get by with less ram air, which may lead to slightly lower aerodynamic drag. Drag is defined as the power necessary to turn the fan. However, the engine cooling fan must be larger or

This modified 1957 Chevrolet Bel Air has a typical mechanical flex-type cooling fan. This fan is driven by a V-belt and pulley on the front of the water pump, which is driven by the engine crankshaft.

more powerful to compensate. Sometimes more than one fan is used.

A fan is mounted in front of or behind the radiator to increase the airflow through the radiator core. Most cooling fans in the past had a fan blade on a pulley that was mounted on the same shaft as the water pump impeller. It was driven by a V-belt. This design does not work too well on front-wheel-drive vehicles equipped with a transverse engine. The electrical fan-and-motor assembly was created to handle this type of engine configuration. The fan is mounted in a shroud attached to the radiator and air-conditioning condenser.

Starting in about 1980, the early electric fan systems were controlled

Here is an example of an electric cooling fan encased in its own shroud and mounted in a modified Model A Ford hot rod.

frame, and blades of either plastic or steel are riveted to the frame. These blades will flatten out from their curved shape at a preset RPM to reduce aerodynamic drag on the engine. The more drag produced by the fan, the greater the reduction of available horsepower. Engine-mounted fans are less effective in traffic because they do not pull much air when the engine RPM is low.

The biggest advantage of flex fans is their ability to pull more air through the radiator at idle than a viscous clutch fan. This is helpful on engines running high compression ratios that are more difficult to cool. The flex fan aids in idle or slow speed operation due to the curvature of the blades scooping air; however, they are somewhat noisy at low engine speeds. As the speed of the fan and engine increases, the resistance of the air flattens the blades slightly, so the fan requires less horsepower to turn than a rigid-blade fan. The flex fan does reduce the amount of air

directly by a thermostatic device, such as a coolant temperature sensor mounted in the radiator, cylinder head, intake manifold, or engine block. When the sensor detected heat above a certain temperature via a wax pellet expansion, it turned on the cooling fan. The fan would run only until the coolant temperature was reduced to the calibrated level. The electric fans were connected to their own power source and not affected by the ignition system. If the sensor detected a high temperature, the fan turned on whether the engine was on or off.

Current OEM vehicles are controlled using mostly an engine control module (ECM) to operate the cooling fan or fans. The ECM uses a relay, the engine coolant temperature (ECT) sensor, and maps stored in the computer's memory.

Flex Fan

Older cooling fans have rigid blades that work quite well on electric fans that run at a constant speed. However, on conventional belt-driven fans, they are noisy and

absorb a measurable amount of engine horsepower at higher speeds. One solution is to use a viscous clutch fan (discussed later in this chapter). An alternative design to increase fuel economy, decrease noise, and provide more airflow at low RPM when in traffic is the flex fan.

The flex fans uses a stainless steel

This older modified car uses a red flex fan that provides more airflow at idle.

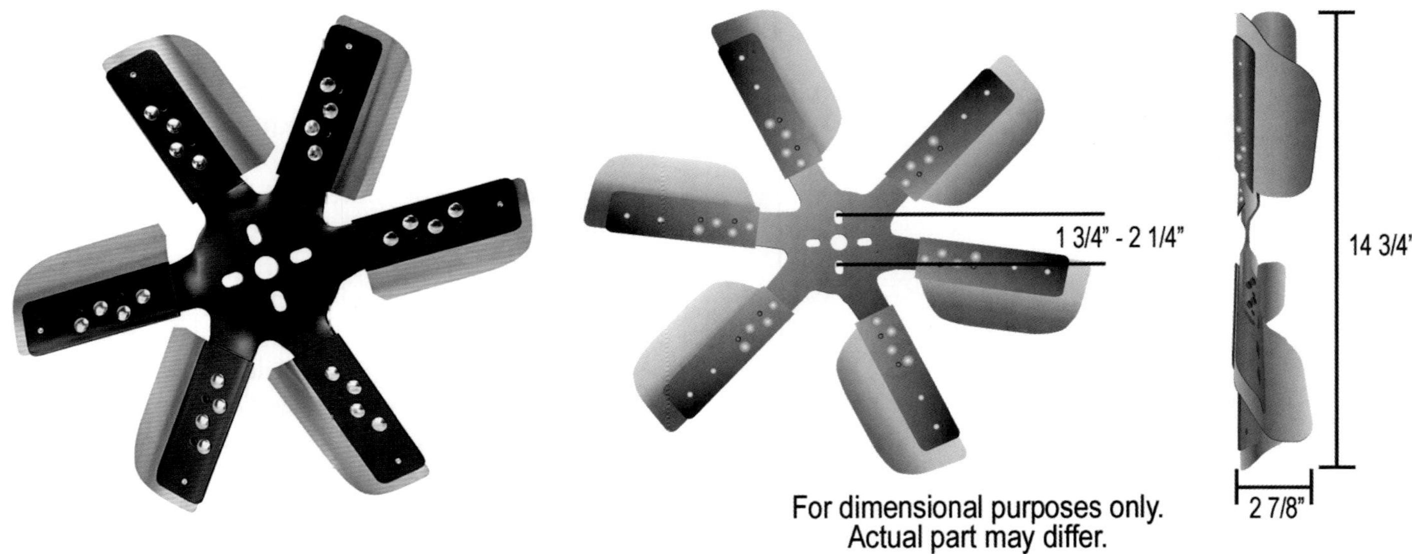

A flex fan can provide an increased amount of airflow in traffic. The fan blades move, or flex, with each rotation, and the blades flatten out when the vehicle itself is moving fast enough to maintain a strong air stream through the radiator. (Photo Courtesy Derale Performance)

the fan can move with each rotation, usually both the fan and the vehicle itself are moving fast enough to maintain a strong air stream through the radiator.

Modified Stock and Race Car Cooling Fan Issues

The most-common modified car or race car cooling system problems concern low-speed cooling, which can be attributed to low airflow through the radiator among other issues. If the radiator is sized correctly and there is a high-flow water pump and a larger thermostat opening, increasing airflow at low vehicle speeds should remove sufficient heat from the radiator to keep the engine at a controllable temperature.

Engine-driven fans can move a tremendous amount of air, but they are also compromised by slow engine speed at idle while delivering sufficient airflow at higher engine speeds. Electric fans have become the norm

with Ford, General Motors, Toyota, Honda, and Fiat-Chrysler because they can move enough air at low speeds to keep the engine cool. They rely on vehicle speed to push air through the radiator at road speed, which reduces parasitic horsepower losses on the highway.

Several car magazines have reported dynamometer testing showing losses of 35 bhp at peak horsepower from a simple one-piece, plastic, engine-driven mechanical fan. Viscous clutch fans lost between 8 and 19 bhp, depending on the clutch model. A Flex-a-lite Black Magic electric fan driven by the alternator cost 1 bhp.

Multiple Fans

Some modified stock and race car builders believe more cooling fans are better. This is not completely true. You don't really need a fan both behind and in front of the radiator. Preferably, you will use a cooling fan behind the radiator that provides

cooling capacity based on coolant temperature. If your vehicle needs two cooling fans, there are other issues in the design.

Fan Spacing and Fan Shrouds

One problem that is overlooked by modified stock and racing vehicle builders is fan spacing and shrouding. Engine cooling fans need to be shrouded for proper vectoring of air velocity through the radiator. I recommend that you take a look at the OEM version of the application being modified if it is somewhat current. You would not want to research a 1932 Ford that you are putting in a 460-ci engine.

Depending on the position of your fan in relation to the engine, you can have either an axial flow or radial flow cooling fan. An axial flow cooling fan has the fan blades inside the shroud with little or no fan blade tip exposure. Air being drawn through the radiator by the fan is flowing in an axial path along

Good cooling shrouding on a mechanical cooling fan assembly can be seen in this modified car.

the axis of the fan and water pump pulley drive shaft.

In a radial flow cooling fan, the fan either does not use a shroud at all, the fan blades are not enclosed within the shroud, or the tips of the fan blades are just outside the shroud. When this design is used, it becomes a matter of the fan spacing. The best design for maximum airflow through the radiator and cooling is to extend the shroud so it covers the fan or to use spacers to keep the fan blades inside the existing shroud. Most radial cooling fan setups are a byproduct of backyard DIY installations. I have not seen an OEM radial fan system.

Fan Spacing Calculation

The shroud fan position or measurement usually depends on the amount of radial airflow. The higher the radial airflow, the greater this dimension needs to be. For example, a 100-percent radial flow fan will be out of the fan shroud. By contrast, a 100-percent axial flow cooling fan will be completely inside or enclosed in the shroud.

Most engine compartments will allow for a 50- to 60-percent immersion of the fan width, which will produce the greatest amount of airflow. This clearance amount is controlled by the amount of engine movement.

Illustrated here is an axial flow cooling fan where the entire fan blade assembly is inside the fan shroud. This allows the air to flow along the axis of the fan and water pump drive shaft in an axial path.

Here is a radial flow cooling fan where the air being drawn through the radiator is flowing radially away from the engine and not over it.

An aftermarket brand viscous clutch fan is installed on this modified Chevrolet Corvette. This vehicle would have been equipped with a viscous clutch fan from the factory.

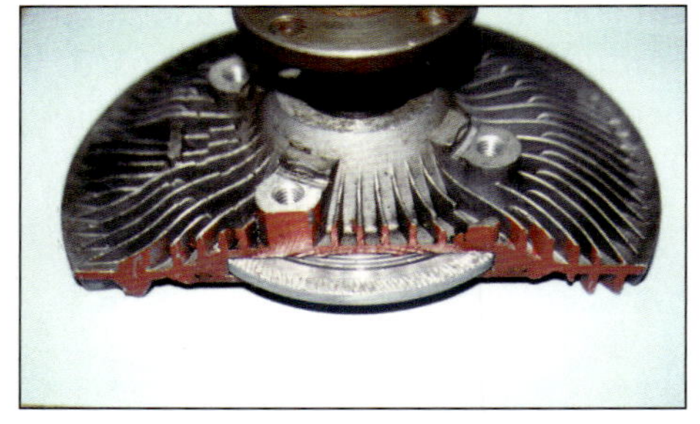

These viscous clutch assembly cutaways show the two viscous members: a drive member (left) and a driven member (right). On the drive member, the clutch plate is splined to the pulley shaft. The driven member is integral with the drive housing, which is bolted to the fan blades. (Photo Courtesy Jim Halderman [left] and Derale Performance [right])

A large clearance of approximately 1 inch will reduce the effectiveness of the fan shroud. The clearance should be as small as possible. Normal OEM fan spacing for mechanical fans was 0.5 inch. Some higher-horsepower engines with large amounts of torque will have more movement and require more clearance from fan tip to shroud. This may be a moot point because with higher-horsepower engines, I use an electric fan.

Viscous Clutch

One way to avoid power loss and excess noise on fan assemblies is by using a viscous clutch fan. When additional airflow is needed, the clutch engages the fan blades to provide maximum cooling. When maximum cooling is not needed, the fan "slips" to prevent wasting engine power. The fan clutch uses a silicone-fluid coupling, somewhat like an automatic transmission torque converter.

There are two viscous members: a drive member and a driven member. The clutch plate is splined to the pulley shaft on the drive member. The driven member is integral with the drive housing, which is bolted to the fan blades. Both of these members have circular grooves that are closely mated to each other.

Silicone viscous coupling fluid is stored in a reservoir chamber in front of the pump plate. A bimetallic spring controls the flow of fluid using a valve attached to the spring. This bimetallic spring senses the air temperature directly behind the radiator to engage or disengage the viscous clutch as needed.

When the radiator air stream is cold, the silicone fluid remains trapped in the reservoir. With little or no fluid in the working chamber, the clutch members are uncoupled

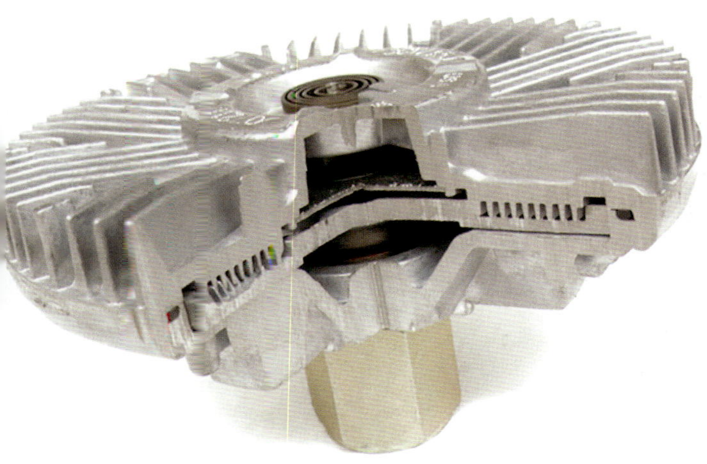

The bimetallic temperature-sensor spring is attached to a valve on a viscous clutch cooling fan assembly. When the bimetallic spring cools down, the fan becomes uncoupled and freewheels. (Photo Courtesy Derale Performance)

If you have a noisy engine, it may be because the fan is engaged all the time. A quick and easy test is to use a screwdriver to disengage the spring from its seat. If the viscous clutch disengages, the clutch is bad and needs to be replaced. Here is front view of viscous clutch bimetallic spring. (Photo Courtesy James Halderman)

and there is no fan rotation. As the bimetallic spring gets hot (around 165 to 185°F), it moves an arm to uncover an opening in the pump plate, letting the silicone fluid flow into the working chamber. The fluid viscosity causes the members to couple and drive the cooling fan whenever there is enough fluid to fill the spaces between the grooves.

When the viscous clutch is coupled, the fluid continuously circulates between the reservoir and the working chamber. It is again trapped in the reservoir when the bimetallic spring cools down the fan, becomes uncoupled, and freewheels.

Electric Fans

Most modified stock and race car builders rarely consider that the early 1960s and 1970s Delco, Mopar, and Ford alternators were rated at 60 to 70 amps. They were not designed to crank out maximum amperage at

Champion Cooling Systems' 16-inch electric cooling fan includes a shroud and shield. (Photo Courtesy Champion Cooling Systems)

A 2008 Chevrolet Impala with a transverse-mounted 3.5L V-6 engine using dual cooling fans is shown.

idle. Late-model alternators were designed to generate greater amperage at idle with Delta-wound stators or 12 pole Y-wound stators. These more-efficient alternators are capable of delivering the 40 amps or

more that are required by dual fans running along with a big electric fuel pump, lights, and an infotainment system. Add the draw from a pair of headlights and perhaps a defroster or air-conditioner clutch/fan, and a load of 50 to 60 amps from the alternator at idle is not unusual. This will also require large 8- to 10-gauge wiring from the alternator to the underhood power source for your fans and multiple solid-ground circuits between the engine and the chassis.

Puller and Pusher Fans

Some modified stock and race car builders believe more cooling fans are better. Yet this is not completely true. It is preferable to have a cooling fan behind the radiator that provides cooling capacity based on coolant temperature. If your vehicle needs two cooling fans, there are other issues in your design. Chapter 5 describes how air flows through the radiator, showing why the position of the fan behind the radiator is best.

The world of electric cooling fans contains two types: the puller and the pusher. Puller fans mount on the engine side (back) of the radiator core and pull air through the core. The fresh airflow is pulled through the grille, then through the core, and exits out the back side of the radiator. Puller fans are the best

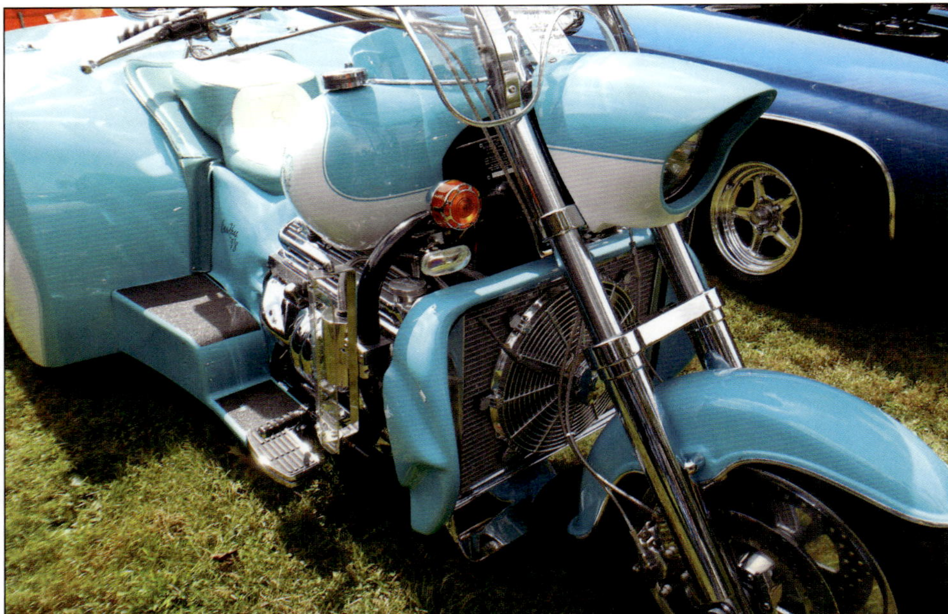

This modified tricycle uses a Chevrolet small-block V-8 engine. With limited space due to the design, there is no room for a puller cooling fan behind the radiator. In this case, the builder used a pusher fan located upstream in front of the radiator.

and most common way of mounting a fan when space is not a limitation between the front of the engine and the back of the radiator core. All mechanical-engine pulley-driven cooling fans are the puller type.

Pusher fans mount on the grille side (front) of the radiator core and push air through the radiator core. Fresh airflow enters through the grille, is pushed through the radiator core, and exits out the back side of the radiator. Pusher fans are used usually where there is a limited space for a puller fan. Pusher fans that are mounted on the top of a radiator

On the left is a standard blade puller-type electric cooling fan assembly with a shroud. On the right is a scroll blade electric puller-type cooling fan with a built-in shroud.

Alternators

The Delco (now Delphi) 10-DN series alternator used an external electromagnetic voltage regulator. Six individual diodes were mounted in the rear housing with a capacitor for protection. A 14-pole rotor and Y-type stator provide current output. Field current is drawn from rectified output and travels through a B-circuit. The terminals on a 10-DN are labeled BAT, GRD, R, and F. DN series was used from 1963 to 1986 and was a Y-stator design with 60- to 70-amp output at full load but not at idle.

The Delphi SI series was the first GM alternator with a Delta wound stator capable of delivering 40 amps at idle or start-up. Most SI models have a rectifier bridge that contains all six rectifying diodes. The regulator is a fully enclosed unit attached by screws to the housing. Field current is drawn from unrectified alternating current (AC) generator (alternator) output and rectified by an additional diode trio.

Delco Remy designed the AD alternator charging system used on GM vehicles, which monitors voltage with a computer. The AD200 series alternators or AC generators went into production in 1999. The AD200 designation refers to second-generation (200), air-cooled (A), dual (D) internal fans. There are three different models depending on unit diameters, and they have amperage output ranges from 102 to 150 amps. The AD alternator uses an offset-wound stator for more consistent output voltage.

Motorcraft, a division of Ford Motor Company, makes most of the alternators used on domestic Ford vehicles. Model and current rating identifications for later models are stamped on the front housing with a color code. Motorcraft AC generators prior to 1985 are used with either an electromechanical voltage regulator or a remotely mounted solid-state regulator.

The Motorcraft JAR AC generators are rated at 40 to 80 amps. The sealed rectifier assembly is attached to the slip ring-end housing. On early models, the connecting terminals (BAT and STA) protruded from the side of the alternator in a plastic housing. Current models use a single pin stator (STA) connector and separate output stud (BAT). The brushes are attached to and removed with the regulator. A Y-type stator is used with a 12-pole rotor.

Chrysler manufactured all of the alternators for its vehicles built in the United States until the late 1980s, when it phased in Bosch and Nippondenso alternators. Chrysler was the first US manufacturer to use an alternator or AC generator in the late 1950s. Chrysler used two alternator designs from 1972 to 1984. The standard-duty alternator, rated from 50 to 65 amps, is identified by an internal cooling fan and the stator core extension between the housings. The heavy-duty 100-amp alternator has an external fan and a totally enclosed stator core. Identification also is stamped on a color-coded tag on the housing.

Chrysler standard-duty alternator used a Y-type stator connected to six diodes. Although both brush holders are insulated from the housing, one is indirectly grounded through the negative diode plate, making it a B-circuit. The 100-amp AC generator had a delta-type stator. ■

The new and smaller Delphi CS-series alternators (far right) were used from 1986 to 1999 on GM vehicles. They produce current outputs similar to larger alternators. This series includes models CS-121, CS-130, and CS-l44. CS stands for charging system. All SI models (far left) have a capacitor installed in the rear housing to protect the diodes from sudden voltage surges and to filter out voltage ripples that could produce electromagnetic interference (EMI). Voltage is adjusted by removing the adjustment cap, rotating it until the desired setting (low, medium, medium-high, or high) is opposite the arrow on the housing, and reinstalling it in the new position. The two alternators in the center are for truck applications.

Champion Cooling supplies an electric cooling fan kit that contains all of the components needed for the typical installation of a cooling fan on a modified stock or racing vehicle. (Photo Courtesy Champion Cooling Systems)

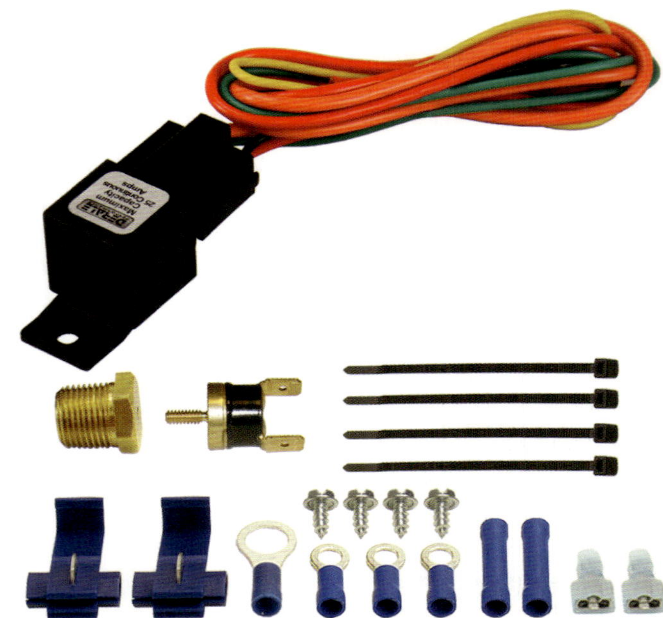

Derale offers a cooling fan kit that includes a two terminal temperature probes and a brass fitting along with the control relay, terminals, wire ties, and instructions. (Photo Courtesy Derale Performance)

core, which is mounted horizontally or flat, will pull the fresh airflow in and then push the air down through the core and exit out the bottom. In your design, make sure there is enough air space below the radiator (approximately the thickness of the radiator core) to remove the warm air out of the core.

Most companies that sell electric cooling fans also offer kits with wiring, instructions, and relays. A good ground also means the ground wires should be of equal size as the power leads. The biggest electric fan won't run at anywhere near peak efficiency if the ground circuit suffers from resistance.

A simple voltage drop test will tell you if the wiring circuit is an issue. Robert Bosch quotes the 3 percent rule; that is, the voltage drop cannot be greater than the applied voltage. This means in a 12-volt system, a 3-percent drop would be 0.36 volts, so you are only allowed a 0.36 voltage drop across the ground.

Electric Cooling Fan Control

Fan control is the management of the rotational speed of an electric cooling fan. Different types of electric cooling fans are used to provide adequate engine cooling, and different fan control mechanisms balance their cooling capacities and noise.

Thermal Sensor–Controlled Cooling Fans

Thermal sensor–controlled cooling fans use a sensor or thermostatic switch to determine when to increase or decrease the speed of the fan. This regulates the temperature to keep the engine from overheating. If your modified stock vehicle is a late model with an engine management computer, the electric fan is most likely controlled by an ECM using temperature information from the ECT sensor.

When you build and wire your electric cooling fan, there are several different approaches to powering and operating the fan. You can use a thermostatic switch or sensor located in the intake manifold or cylinder head and a sensor, relay, and wiring kit from one of the cooling fan manufacturers.

Switch-Controlled Electric Cooling Fans

The simplest form of electric coolant fan control is to install a simple switch kit. This is a dash-mounted switch that can be turned on when in traffic or if the temperature gauge shows an increase in cooling system temperature.

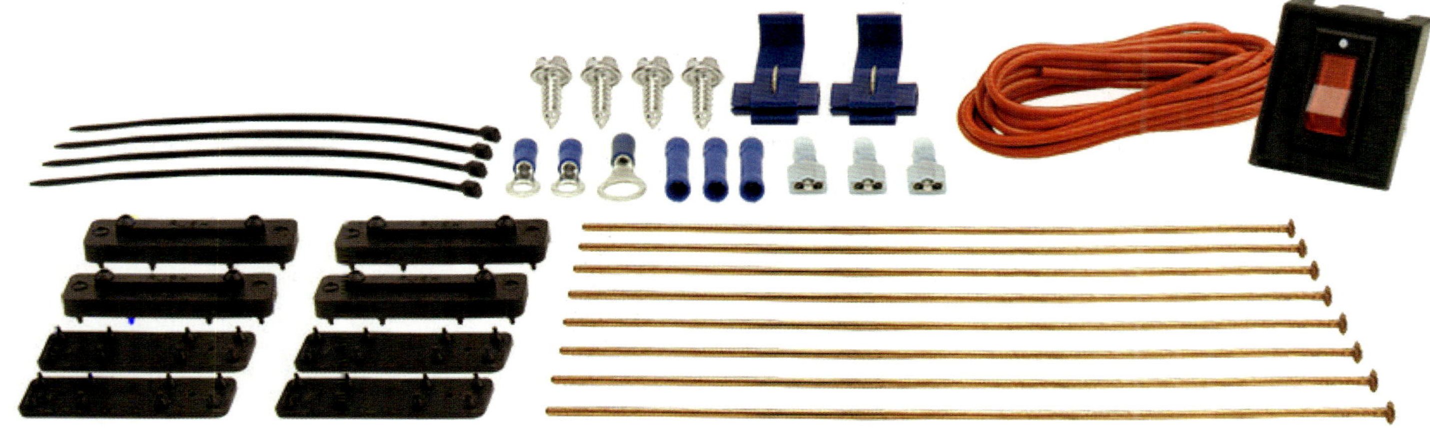

Derale also has a simple electric cooling fan manual switch to turn the cooling fan on or off. (Photo Courtesy Derale Performance)

PWM Controller–Controlled Cooling Fan

The pulse width modulated (PWM) controller is another form of cooling fan control when you are not using the manufacturer's ECM to control the fans. Cooling fans run at 100 percent, even when less airflow is needed. They are constantly turning on and off as the engine temperature changes. Reaction-based temperature control systems using just a temperature sensor in a cooling passage can cause amperage spikes, noise, and premature wear. The Derale PWM fan controller monitors the engine's temperature to engage the cooling fan(s) controlling the engine temperature. The PWM helps increase cooling efficiency and reduces the amperage needed to operate the fan(s).

Cooling Fan Selection

The toughest question when choosing an electric cooling fan is how do I select the one that bests fits my needs? There are a lot of electric cooling fans in the marketplace, and unfortunately, there is no one correct answer. My research had uncovered a few tips that might be helpful.

Things to consider:

- Straight-blade fans move more air than curved-blade fans but will be a lot nosier.
- There is not an industry standard for electric fan rating. Most of the manufacturers use a cubic feet per minute (CFM) rating, often expressed in freeflow and not when placed behind a radiator. This makes comparisons of electric fans difficult.
- Any fan's highest CFM rating occurs with no airflow restriction in front of the fan. This restriction value is listed in inches of water. As the restriction increases due to a thicker radiator core, flow volume drops while current flow increases slightly.
- One rule of thumb that I found in my research is there are 10 amps of current flow per 1,000 cfm of airflow. This is not precise for all cases, but if a fan is rated at 3,000 cfm and only requires 10 amps, the CFM rating may be hopeful.
- Twin fans can be a plus in tight-clearance situations since staggering the two fans moves them away from the water pump pulley.
- Two fans usually can cover more radiator surface area than one large fan, which makes the twin-fan systems generally more efficient.
- Twin-fan performance is also often enhanced by built-in shrouds that pull air in from the entire core surface as opposed to just the area of the radiator covered by the fan.
- There are two measurements that would be very helpful when selecting a cooling fan for your project car: airflow in cubic feet per minute (CFM) and total system airflow resistance. These

Shown here is Derale's pulse width modulated (PWM) electric fan controller. (Photo Courtesy Derale Performance)

Pulse Width Modulation

The PWM cooling fan controller monitors engine temperature at the radiator and provides the most-accurate coolant flow temperature reading for a cooling system. This type of signal is often used for changing the operation of a cooling fan solenoid to turn the fan off and on. Using Soft Start Technology, fans ramp up slowly, which eliminates harmful ampere spikes.

Electric cooling fans only run between 40 to 60 percent of the time to keep your engine cool. The PWM unit will continuously monitor your engine's temperature and operate the fans at the optimum fan rate as needed between 1 to 100 percent duty cycle. The amount of on-time is compared to 100 percent to give the duty cycle percentage.

PWM controllers result in reduced fan speeds, which means reduced fan noise to compete with the sweet sound of that engine. There are additional features as well. The main features of the controller are:

- Controls fan speed from 0 to 100 percent
- Constantly adjust fan speed to maintain the desired temperature
- Controls multiple electric fans up to 65 amp capacity
- Has a finned aluminum housing
- Includes automatically resetting circuit breaker
- Has a built-in air-conditioning override circuit

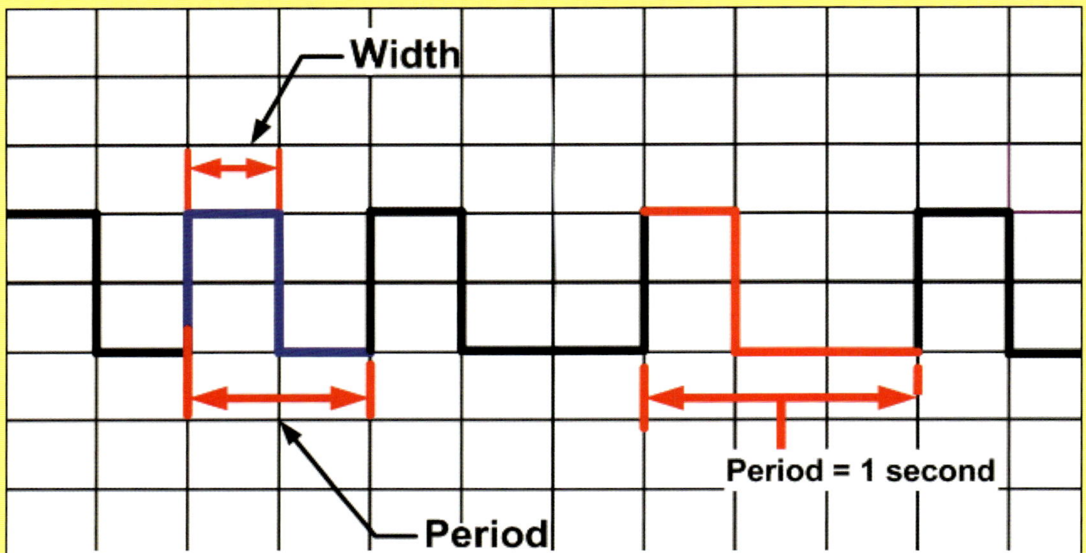

*The term **square wave** is commonly misused to describe a PWM waveform. PWM means the on-time is different and usually changing with respect to the off-time. The pulse waveform has a leading edge, which is referred to as the rise. The on-time is referred to as the duration, and the falling or trailing edge is referred to as the decay. This PWM waveform shown as a 50 percent on and 50 percent off duty cycle. The duty cycle is calculated by dividing the waveform width by the period of time.*

values are determined by the amount of heat to be removed and the air temperature. Manufacturers determine these values from testing that you cannot do, so you get them from the fan manufacturer.

- If you're experiencing overheating and the rest of the cooling system is optimized, increasing airflow with a pair of smaller fans covering the entire radiator core will generally improve airflow and cooling efficiency.

High-performance high-output electric cooling fans incorporated into an engineered shroud can fix the most difficult cooling conditions. Many of these units have a non-finished aluminum shroud that allows the builder to polish, anodize, paint, or simply

This dual electric fan setup from Derale Performance measures about 25x16 for a radiator designed to fit the dimensions of this fan setup. (Photo Courtesy Derale Performance)

Shown here is the single scroll blade–design cooling fan kit with wiring and manual switch. It will increase air turbulence across the radiator. (Photo Courtesy Derale Performance)

BeCool offers a single electric cooling fan mounted on an aluminum crossflow radiator. The kit includes the radiator, cooling fan, wiring, relay, temperature probe, and surge tank. (Photo Courtesy BeCool Performance)

A high-output dual cooling fan set may be a direct fit for a properly sized radiator. Shown here with an aluminum shroud measuring 26x18 inches top to bottom and is only 4 inches in overall depth to offer an adjustable yet precision fit. The dual 12-inch high output electric fans are designed to prosper on thick, tight fin radiator cores. They are offered in a 4,000 cfm airflow. (Photo Courtesy Derale Performance)

leave it raw to create a truly custom look.

Mechanical Fan Pulley Ratio

When choosing a mechanical fan or viscous clutch fan, the pulley ratio is determined by the engine accessory drive components. Fan horsepower and air-conditioner compressors are the highest contributors to the total accessory belt load. The pulley diameter is sized comparative to the type of drive belt, either V-belt or serpentine. Pulley diameter and thus pulley ratio also impacts the water pump output.

Most high-performance pump companies recommend a crankshaft to water pump pulley ratio of 1:1.4 to 1:1.5 for street engines that traditionally run primarily at low engine speeds, so this would be a good compromise for your street rod. Some of them have a pulley ratio that may run as high as 10,000 rpm, which might present a problem for a viscous clutch because it may cause excessive slip. The 1992 to 1996 Chevrolet Corvette LT1 engine used a camshaft drive water pump that ran at a crankshaft to water pump ratio of 1:1.

Wiring

You will need to wire the cooling fan to meet the needs of your project vehicle. I have included a typical wiring diagram for both a single electric cooling fan and a dual fan setup that is a good starting point. It is best if you obtain the wiring diagram that matches the fan you have chosen.

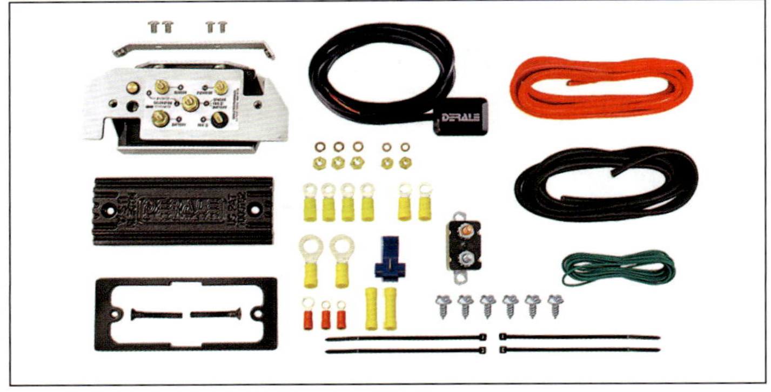

This installation kit contains all of the connection components for a PWM-controlled cooling fan, including the controller. (Photo Courtesy Derale Performance)

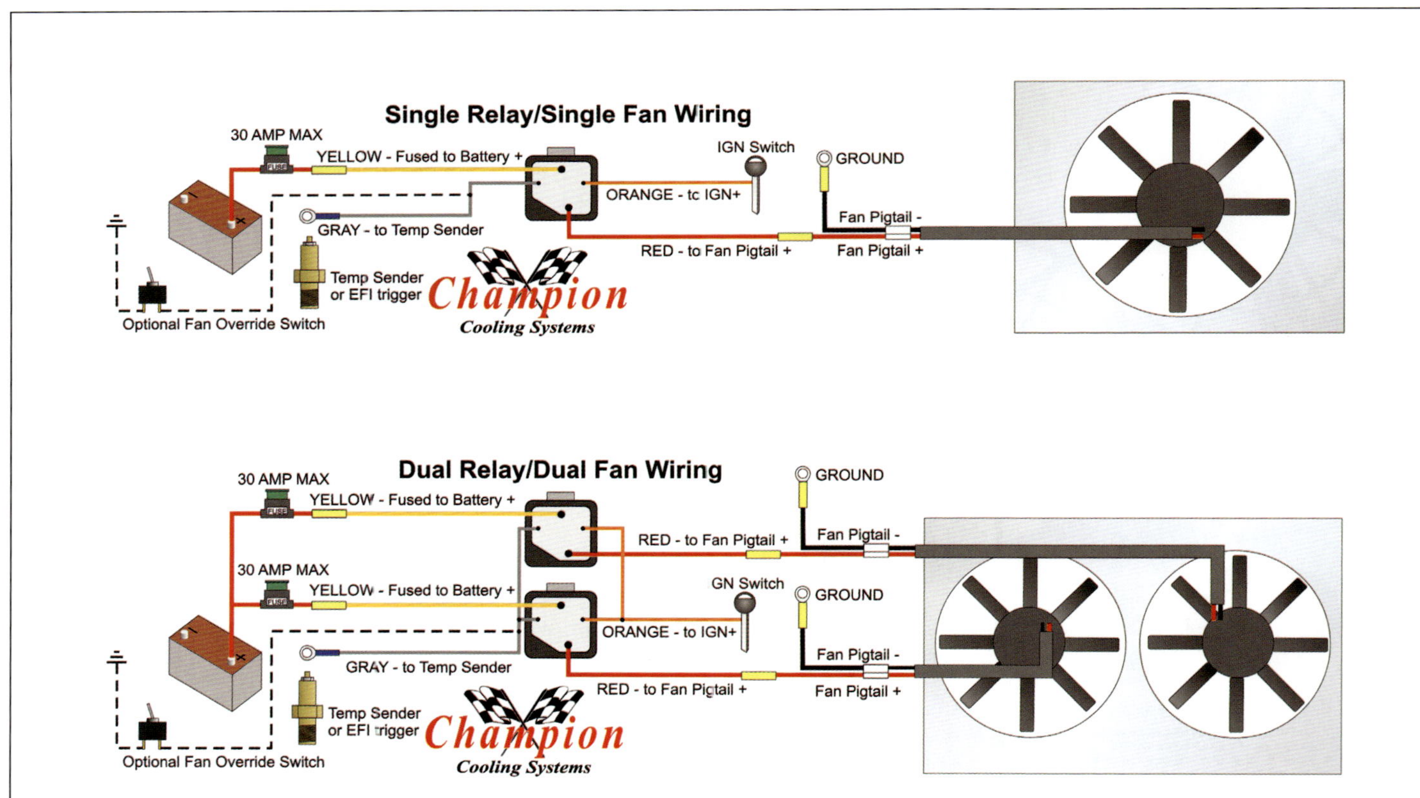

A typical setup for wiring either a single electric cooling fan or a dual electric cooling fan for your project vehicle is shown. It operates using a temperature sensor installed in a hot coolant passage. (Photo Courtesy Champion Cooling Systems)

*B*ASIC *S*YSTEM *D*IAGNOSIS

The goal of any diagnostic strategy is to make a plan for each specific diagnostic situation. That way, you can diagnose and repair your cooling system efficiently. This process examines the engine cooling concern or condition (what the vehicle is doing wrong) along with visual inspections, quick checks, and the vehicle's self-diagnostic ability. The symptom diagnosis codes are then found in online service information sites such as Shop-Key Pro and Alldata.

The steps in a diagnostic system check are:

- Verify what the concern, condition, or problem is before proceeding.
- Locate your cooling system operation description in order to familiarize yourself with the system functions. You can use the online service information websites or the factory shop manual. Get the cooling fan's description and operation and the cooling system's description and operation.
- Perform a visual and a physical inspection. These inspection are extremely important and can lead to correcting a condition

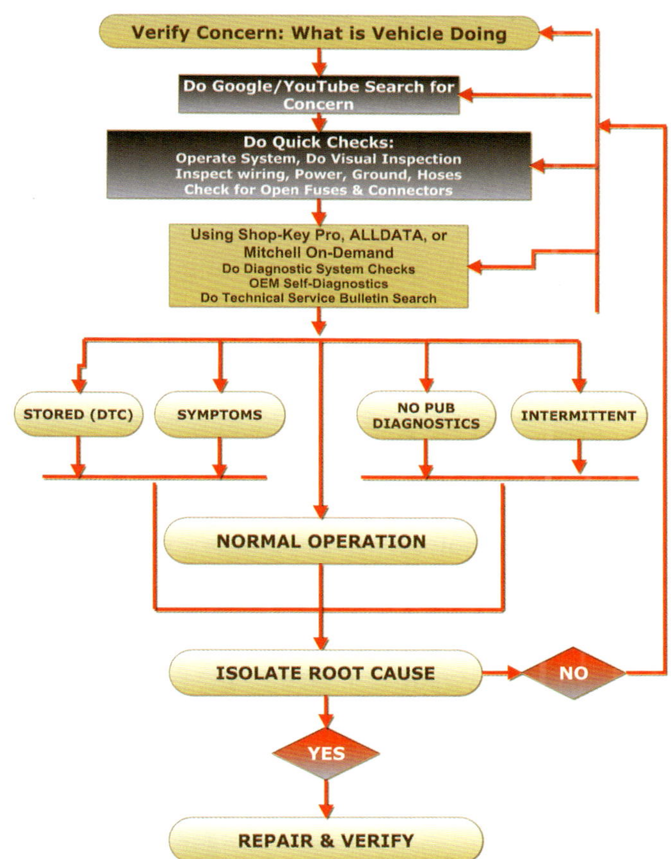

This diagnostic flow chart is a troubleshooting process that takes into account the self-diagnostic capabilities and the service information provided by the original manufacturers. General Motors calls this diagnostic approach strategy-based diagnosis, which is a scientific process of elimination. The acronym DTC in the flow chart stands for diagnostic trouble code, which is a two-character flash code on pre-1996 vehicles or a five-character codes, such as P0116 for the engine coolant sensor, after 1996. These codes can only be read with a scan tool.

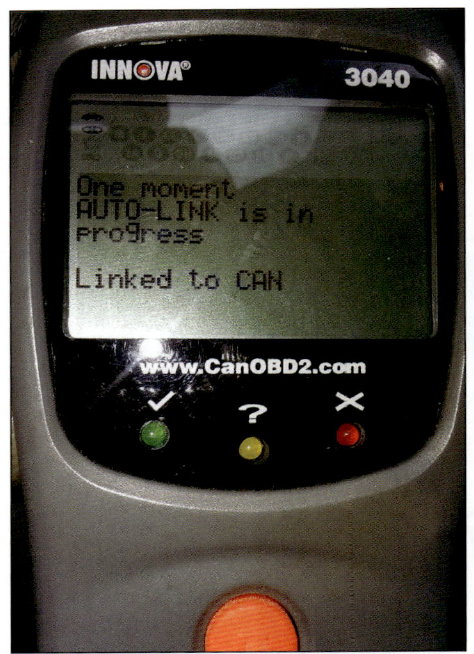

The Innova scan tool code reader will tell you the correct engine operating temperature; if there are any DTCs stored, pending, or active; and other useful engine information. The scan tool shown does not have bidirectional capabilities.

The Snap-On Modis scan tool is a common sight at many professional dealerships. Snap-On tools are some of the best in the industry.

without testing. It may also help reveal the cause of an intermittent condition.

- Perform a diagnostic check. Use a scan tool if you have one. They can be purchased or rented from an auto parts store. Scan tool data will tell the correct engine operating temperature, if there are any DTCs for the vehicle, and other useful engine information. The scan tool will also check that the engine control module (ECM) or powertrain control module (PCM) and malfunction indicator lamp (MIL) are operating correctly.
- Locate the correct cooling system symptom from the online service information. Perform the tests and inspections associated with the symptom.
- Repair the problem and verify it is fixed.

Information Gathering

For an accurate diagnosis, you need to know how your cooling system operates normally. This will help you determine whether there really is a problem. When you operate the system, you might have to check the owner's manual or the online service information.

System Description

The typical engine cooling fan system is composed of one or two electric cooling fans and a cooling fan control module. The ECM or PCM

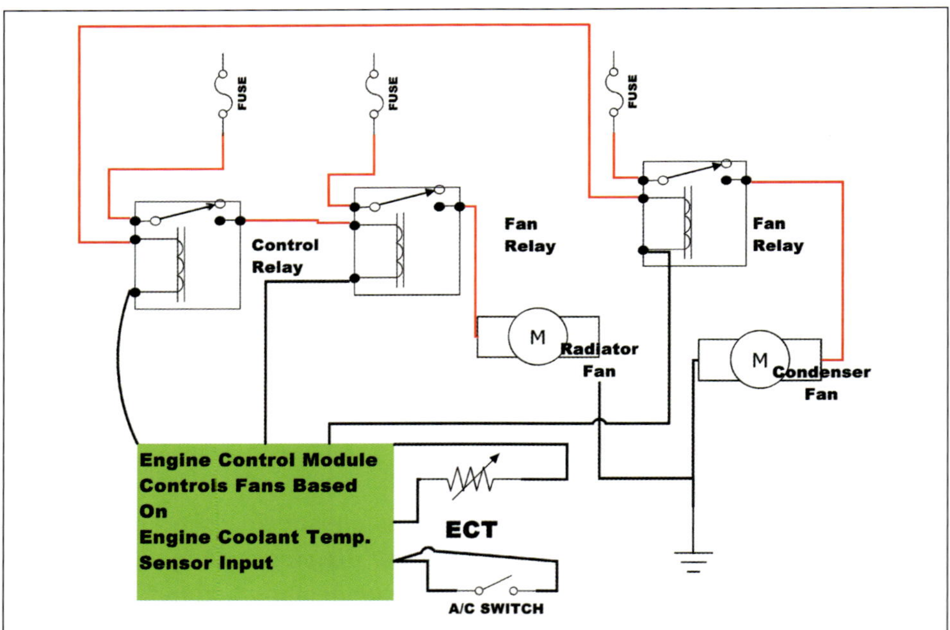

An electric cooling fan circuit is operated by the engine management computer, in this case an ECM. Based upon input from the ECT sensor, which is a thermistor (the hotter it gets the less resistance), the ECM will ground either one or both of the fans relays to turn on the cooling fans.

controls the fan speed by sending a pulse width modulated (PWM) signal to the cooling fan control module. The cooling fan control module varies the voltage drop across the cooling fan motor in relation to the PWM signal, which enables cooling fan operation at variable speeds.

The ECM or PCM will only operate the cooling fan at certain percentages to prevent undesirable noise and vibration. The cooling fan control module is thermally protected to prevent module damage in the case of a short circuit condition in the cooling fan motor. On past systems or aftermarket electric cooling fan kits for modified vehicles without computer-controlled fuel systems, an electric cooling fan will be operated by a sensor in an engine coolant passage or a simple switch. When the sensor reaches the calibrated temperature, the coolant fan will turn on.

Some aftermarket cooling system electric fans can come with either a simple on/off switch or a relay that is controlled by a wax motor temperature sensor that you install in a coolant passage. This temperature sensor is installed in a coolant passage in the intake manifold or cylinder head.

Service Manuals

To find information specific to your system, look at its service manual. Service manuals can be found at several online sources, including Real Auto Service Manuals at realism.com. You can also download PDF versions of service manuals from youfixcars.com/auto-service-repair-manual. A Google search will reveal that there is a lot of material available online.

YouTube

There are many YouTube channels providing videos of repairs being performed, and some of these can be quite good. However, it is best to proceed with caution. There is no vetting process, so anyone can post a video there. Use unvetted information sources with caution. The YouTube channels Scanner Danner and Eric the Car Guy are very informative.

Inspections

After you have verified your concern or problem and gathered information, perform a visual inspection. Several of the symptom procedures call for a careful visual and physical inspection. This can lead to correcting a condition without further tests and can save time. Check for any unusual smells, noises, or movements. Look at your maintenance records to see what work has been done before.

Quick Check

Depending on how complicated the system is, the quick checks you should do include:

- Operate the engine and bring it up to operating temperature.
- Inspect the wiring harness, power, and ground.
- Check for blown fuses and separated connectors.
- Inspect the fan and sensor connectors (including terminals) for damage and tightness.
- Notice any unusual noises, smells, vibrations, or other movements.

Cooling System Checks

If all of the quick check items are okay, check the following:

- Check the water pump drive belt for tension.

Begin a visual inspection of the cooling system underneath the hood. Look for signs of leaking coolant along the soft radiator hoses and coolant leaking out of the water pump weep hole. If you see smoke rising, it is most likely from a blown cooling system rubber hose.

Use an inexpensive digital multimeter (DMM) to check for battery voltage to the ECM or PCM. Also check that it is properly grounded and to see if the voltage drops across the ground. There should be no more than 0.2 volts drop across the ground.

- Inspect hoses and pipes for splits, kinks, improper connections, leaks, or restrictions.
- Look for coolant overflow or loss.
- Check the surge tank coolant level.
- Check the radiator for corrosion or obstructions.
- Inspect the ECM/PCM harness connectors.

- Inspect the cooling fan electrical center fuse/relay cavities; check for open fuses.
- Check component terminals and harness connectors.
- Make sure that ECM or PCM grounds are clean, tight, and correctly located.
- Use a digital multimeter (DMM) to check for battery voltage to the ECM or PCM and that it is properly grounded (there should be no more than 0.2 volts drop across the ground).
- Inspect for a dirty or restricted radiator or condenser if equipped with an air conditioner.
- Inspect the coolant recovery reservoir for proper coolant level.
- Do a cooling system pressure test.

Coolant Level

Coolant level was included in several of the items in the cooling system checks list. If the level is not correct, you need to fill it to the proper level.

1. Remove the coolant surge tank pressure cap when the cooling system, including the coolant surge tank pressure cap and upper radiator hose, is no longer hot. Turn the pressure cap slowly counterclockwise about one-quarter of a turn. If you hear a hiss, wait for that to stop. This will allow any pressure still left to be vented out the discharge hose.

The expansion tank (left) has plenty of coolant in the tank. This Cadillac ATS uses two tanks: an upper and a lower tank. Expansion or surge tanks provide expansion of your cooling system, providing up to a 1/2 gallon more coolant. This additional coolant is stored and can flow back into the main cooling system when the engine cools down. The radiator pressure cap is directly on top of the expansion tank with no pressure cap on the top of the radiator. Check the coolant level directly at the radiator with the pressure cap removed (left).

In this photo, a pressure tester has been installed in the radiator or expansion tank pressure cap opening. It shows 11 psi of pressure while the engine is running, indicating an internal leak of combustion gas into the cooling system.

A blown head gasket that has been leaking coolant internally from the head and block coolant passage into the cylinder is shown here.

Combustion leak chemical detector and fluid can be used as a chemical test to see if combustion gases are getting into the cooling system from a failed head gasket, cylinder head, or block. With the radiator pressure cap removed, place the tool into the radiator or expansion tank and pour some of the fluid into the radiator or tank. Using an engine vacuum or a Mityvac, suck some engine coolant into the tool. If the solution turns yellow, you have combustion gases in your coolant.

2. Keep turning the pressure cap slowly and remove it.
3. Fill the coolant surge tank with the proper coolant mixture to the indicated level mark.
4. With the coolant surge tank pressure cap off, see if the engine will start. If the engine grunts and you hear the starter solenoid click with no engine rotation, stop immediately. The engine may be hydraulically locked. Skip to checking the Hydra-Lock.
5. If the engine starts and runs, let it run until you can feel the upper radiator hose getting hot. Watch out for the engine cooling fan. By this time, the coolant level inside the coolant surge tank may be lower. If the level is lower, add more of the proper coolant mixture to the coolant surge tank until the level reaches the indicated level mark.
6. Replace the pressure cap. Be sure the pressure cap is hand-tight. Check the level in the coolant surge tank when the cooling system has cooled down. If the coolant is not at the proper level, repeat steps 1 through 4 and reinstall the pressure cap.

Hydra-Lock

Before going any further, you must make sure the engine is not hydraulically locked from a suspected internal coolant leak. A blown head gasket or cracked block or head can cause a loss of coolant going into a cylinder. The coolant could have hydraulically locked the engine.

To check the Hydra-Lock, first see if the battery and cranking motor are in good working order. If the engine cranks and then runs, you need to quickly check for a leaking head gasket. Listed are several methods that can be used to diagnose a head gasket failure on a running engine:

Exhaust Gas Analyzer: Remove the radiator cap on a cold engine and place the 5-gas engine analyzer probe (sniffer) directly above the radiator filler neck or expansion tank. If the hydrocarbons (HC) increases, combustion gases are getting into the cooling system from a blown head gasket.

Chemical Test: Detect combustion gases in the cooling system on a running engine by using a combustion leak chemical detector and fluid. Remove the radiator pressure cap on a cold engine and place the tapered end into the radiator or expansion tank. Remove the top cap, which has a vacuum port on it, and pour some of the blue fluid into the tube. Place the top cap back on. Using engine vacuum or a vacuum pump, suck engine coolant into the tube to mix with the blue fluid. If combustion gases are present, the liquid combination will turn yellow.

Bubbles in Coolant: Remove the water pump drive belt or serpentine belt, then remove the radiator pressure cap on a cold engine. Start the engine. If bubbles appear in the coolant before it begins to boil, the head gasket is blown.

Excessive Exhaust Steam: If there is excessive water, steam, or white smoke coming out the tailpipe, this also can indicate a blown head gasket.

Pressure Testing the Radiator: You can also remove the radiator pressure cap and connect a pressure tester to the radiator or surge/expansion tank pressure cap opening and see if it shows any pressure while the engine is running without the pressure cap installed. If pressure is shown, there is an internal leak of combustion gas into the cooling system.

Rotate the Crankshaft: If the engine just grunts and does not crank with 12.6 battery voltage, check to see if the engine crankshaft rotates. Use a torque wrench and socket to test at the crankshaft balancer. An engine will rotate at 60 ft-pounds on the torque wrench if it is not locked. If the engine will not rotate, remove the spark plugs and rotate the crankshaft with the torque wrench and socket by hand. If coolant blows out the spark plug holes, you have a hydraulically locked engine. This is typically caused by a blown head gasket, cracked block, or cracked cylinder head. You now have to disassemble the engine to see if the head gasket is blown and check for a warped head surface and/or a cracked head or block.

Cylinder Head

Checking for a cracked engine block or cylinder head will mostly likely require you to take that component to a machine shop for testing. Cast-iron heads and blocks are checked with a process called Magnaflux. With Magnaflux, a fine iron powder is sprayed in the areas prone to cracks, such as between the valves. A magnetic bridge is placed over the crack area. If there is a crack, the powder will be attracted to the strong magnetic concentration around it. The magnetic lines of force are conducted through the iron part and will concentrate on the edges of a crack.

With aluminum heads, testing is done with dye-penetrant testing. A dark red penetrating chemical is sprayed on the component and then cleaned up. A white powder is then sprayed over the test area. If a crack is present, the red dye will stain the white powder at the crack area.

A third option is fluorescent-penetrant testing, which requires a black light. It can be used on iron, steel, or aluminum parts. Cracks will show up as bright lines when viewed with a black light.

A last resort is pressure testing. In this test, a machine shop pressure tests the cylinder heads and blocks with air to check for leaks.

Diagnostic System Checks

There are several steps to checking diagnostic systems information. Most online service information websites, including shopkeypro.com, alldata.com, or MitchellOnDemand.com, provide symptom diagnosis for all systems. You could check the operation of the cooling system in an identical vehicle to see if it works differently. In any case, it's important to get as much information about the problem as possible.

Technical Service Bulletins

Automotive manufacturers publish technical service bulletins (TSBs) about problems that have appeared

The top deck surface of a block is tested using magnetic crack inspection equipment, which is basically a large magnet placed over the crack. Magnafluxing is an advanced procedure that uses strong magnetic fields to test the structural integrity of metals, especially iron and iron-based. (Photo Courtesy Jim Halderman)

on certain models. These bulletins are all online through one of the service information providers or directly from the OEM. For example, the GM website is called Si2000.

Sometimes, you can Google a concern and a TSB will pop up. Often, you will find a TSB that exactly describes the problem you're having and tells you how to fix it. Once you have checked for a TSBs, you will now need to select a diagnostic path. There are two paths in a typical diagnostic strategy to choose from: diagnose stored diagnostic trouble code (DTC) or diagnose cooling system symptoms.

Diagnostic Trouble Codes

A diagnostic trouble code (DTC), can indicate an issue in the cooling system. A factory level scan tool can be connected to the data link connector (DLC) on vehicles built after 1996 and non-modified vehicles. The DLC is generally found on the left side under the dash.

Once hooked up with the ignition on, follow the menus to check DTC. If any are present, the malfunction indicator lamp (MIL), also known as the check engine light, will be on. Use the online service information to diagnose the DTC before proceeding.

In order to set certain DTCs, the ECM or PCM must command the cooling fan system to turn on. A delay of about 12 seconds may occur before the cooling fan activates or changes speed when being commanded with an aftermarket factory level scan tool. Using the scan tool to operate the cooling fan will not set certain DTCs. In these cases, run the vehicle until it reaches an appropriate condition for the ECM/PCM to command the cooling fans to turn on.

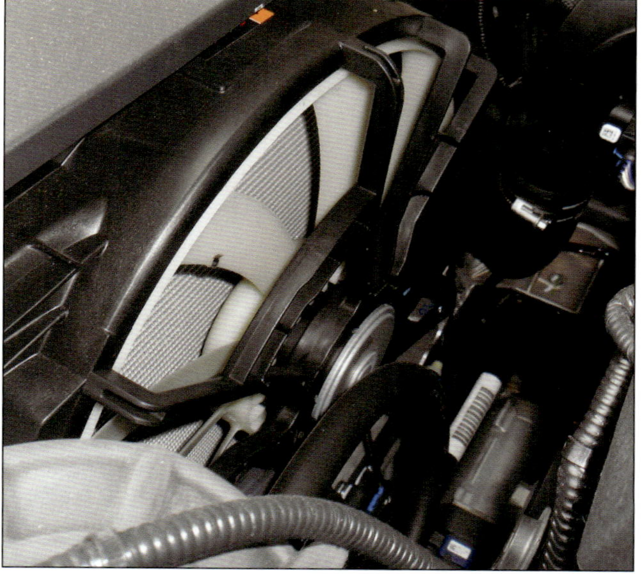

An electric cooling fan was found on this late-model vehicle, whose operation is controlled by the ECM. The ECM or PCM will only operate the cooling fan at certain percentages to prevent undesirable noise and vibration. The cooling fan control module is thermally protected to prevent module damage in the case of a short circuit condition in the cooling fan motor.

Intermittent Electrical Conditions

Many intermittent conditions occur with harness or connector movement due to engine torque, rough roads, vibration, or physical movement of a component. Take a look at the following to help isolate an intermittent condition:

- Water intrusion in connectors, terminals, or components
- Poor connector mating
- Terminal contact
- High circuit or component resistance (High resistance can include any resistance, regardless of the amount, that can interrupt the operation of the component)
- Harnesses that are routed too tight and/or chaffed circuits
- High or low ambient temperatures
- High or low engine coolant temperatures
- High underhood temperatures

Look under the dash near the steering wheel; you should see the data link connector (DLC). This 16-pin DLC is located to the left of the steering column. Although the DLC can also be on the right side of the steering column. Consult the online service information for the exact location for the vehicle you are working on.

- Heat buildup in components or circuits due to circuit resistance, poor terminal contact, or high electrical load
- High or low system voltage
- High vehicle load conditions
- Rough road surface
- Electromagnetic interference (EMI) or circuit interference from relays, solenoids, or other electrical surge
- Incorrect installation of non-factory, aftermarket, or after factory add-on accessories

Isolate Root Cause, Repair, and Verify Fix

Once you've found the problem, repair it following the procedures in the service information. Manufacturers publish step-by-step procedures in their service information for use in diagnosing systems. As you follow the procedures, do each step in order. Don't skip steps or do them out of order because you may diagnose the problem incorrectly and replace the wrong part. If special tools are required, be sure to use them.

Once you have fixed the problem, check the system under the same conditions. You may also have to do some preventive work to make sure the same problem doesn't happen again.

Engine Overheating Diagnostics

The most common concerns with the cooling system is overheating, or running hot due to air or component failure. There are three basic factors that determine engine cooling efficiency: radiator surface area, coolant speed through the system, and the radiator airflow amount. In the

<table>
<tr><td>

Normal Operation

Sometimes a vehicle is working the way it's supposed to but you think there's a problem. In this case, compare your vehicle with an identical one and see if they are running the same. You can also check the owner's manual, the online service information for the vehicle, and the TSBs.

</td></tr>
</table>

repair world, one of the major concerns is air in the system from leaks, so testing for and locating leaks is very important.

The diagnostics list on the next page can help determine the problem.

Note: Extended operation with a low coolant level can cause engine internal component failure. Use caution when diagnosing this issue.

Coolant in the Exhaust System

Condensation in the exhaust system can cause an odorless white smoke during engine warm-up. If present, coolant in the exhaust system creates a distinctive burning odor in the exhaust. To diagnose this, idle the engine at normal operating temperature and look for heavy white smoke coming out of the exhaust pipe.

Does the white smoke have a burning coolant type odor? If yes, you have coolant in the engine oil and in the combustion chamber. You will need to repair the engine internal coolant leak with possible internal engine damage. If no, inspect the coolant recovery system with the engine idling. Does the coolant recovery system discharge coolant into the expansion tank while the engine is idling?

If no, visually inspect the hoses,

pipes, and hose clamps at the following locations: coolant surge tank, water pump, throttle body, heater core, and radiators.

If yes, visually inspect these components:

- Coolant pressure cap: Pressure test the coolant pressure cap. Does the coolant pressure cap hold pressure? If no, replace the cap. If yes, check for other leaks.
- Block heater
- Core plugs
- Cylinder head gaskets
- Engine block
- Intake manifold
- Radiators
- Thermostat housing
- Water pump

Are any of the listed components leaking? If yes, repair or replace the leaking component. If no, pressure

The cooling system can be inspected for leaks by putting dye in the system and using a leak-detection flashlight. The large, awkward lamps that were once the industry standard are a thing of the past. Compact, lightweight flashlights are now the norm, and these are typically battery powered for ease of use. There are several types of leak-detection flashlights available. (Photo Courtesy Jim Halderman)

Steps	Action	Yes	No
1	Check for a loss of coolant. Is there a loss of coolant?	Go to Step 2	Go to Step 3
2	Fill system to specified level. Go to the service information or manual for loss of coolant and locate the problem and repair it. Does the engine overheat?	Go to Step 2	System Okay
3	Check for low coolant protection; go to Coolant Concentration Testing in the service information or manual. Is the coolant to the correct concentration?	Go to Step 5	Go to Step 4
4	Check for a loss of system pressure, go to Cooling System Leak Testing in the service information or manual. Is there a loss of system pressure?	Go to Step 5	Go to Step 6
5	Check for a faulty ECT sensor. If the ECT is faulty, a DTC will be set and MIL will be on. Go to DTC diagnosis in the service information or manual. Is the sensor operating properly?	Go to Step 6	Go to Step 7
6	Check for a cracked coolant recovery reservoir or a leaking hose. Go to Coolant Loss in the service information or manual. Is the reservoir cracked or is a hose leaking?	Go to Step 7	Go to Step 2
7	Repair or install new components or parts as needed, then recheck. Does the engine overheat?	Go to Step 8	System Okay
8	Check for incorrect drive belt tension. Go to Drive Belt Tensioner Replacement in the service information or manual. Is drive belt at the right tension?	Go to Step 9	Go to Step 7
9	Use a scan tool or timing light to check for incorrect engine ignition timing. Is the ignition timing incorrect?	Go to Step 7	Go to Step 10
10	Check for a damaged water pump driveshaft on the 1992 to 1996 Chevrolet LT1 Reverse Coolant Flow Engine. Is the water pump driveshaft damaged or is the seal leaking?	Go to Step 7	Go to Step 11
11	Check for obstructed radiator airflow or bent radiator fins. Is the radiator airflow obstructed?	Go to Step 7	Go to Step 12
12	Are the cooling system passages blocked? Locate the engine coolant flushing procedure in the service information or manual. Add the recommended coolant to your engine. Is there still a block?	Go to Step 3	Go to Step 13
13	Check for inoperative cooling fans, either viscous clutch or electric. Are the cooling fans working?	Go to Step 14	Go to Step 7
14	Check the thermostat by using diagnostic information from the service information or manual. Is the thermostat stuck in the closed position?	Go to Step 15	Go to Step 16
15	Replace the thermostat. Does the engine overheat?	Go to Step 16	System Okay
16	Check for a faulty water pump. Are the impeller blades broken or eroded from cavitation erosion?	Go to Step 7	Go to Step 17
17	Check for a missing and/or damaged radiator upper air deflector and/or center air deflector or air dams. Are the deflectors or air dams missing or damaged?	Go to Step 7	Go to Step 18
18	Check the radiator cooling capacity. Is the proper-sized radiator being used on your vehicle?	Go to Step 2	Go to Step 19
19	Replace the radiator. Is the repair complete?	System Okay	

test the cooling system and visually inspect the components listed above. If you cannot see any leaks, rent a black light leak-checking system, add the dye to the cooling system, and check for signs of the dye indicating a leak.

Pressure Testing a Cooling System

If the engine is not locked, it can be pressure tested for leaks. Use a hand-operated pressure tester to simulate the engine operating pressure listed on the radiator pressure cap. Do not pump the pressure beyond that specified by the manufacturer. Too much pressure in the system may cause the water pump, radiator, heater core, or hoses to fail.

First, remove the radiator cap and select the correct adapter for the vehicle and engine you are working on. Attach the adapter and pressure pump to the top of the radiator or surge/expansion tank. Pump the plunger up to the pressure listed on the radiator pressure cap to pressurize the cooling system.

If the system holds the 15 psi or whatever the specification is, and the system pressure is not holding, you should see leaks at the radiator, tanks, hoses, or heater core. You can also be leaking coolant internally into the combustion chamber from a blown head gasket or crack.

Next, select the correct adapters to test the radiator pressure cap. After selecting the correct adapter, connect it to the tester and connect the pressure cap to the adapter. Use the hand-operated plunger to pressurize the cap up to the pressure listed on the pressure cap. If the cap will not hold pressure, replace it.

Cooling system pressure testers come with an assortment of adapters for a variety of engine applications. This kit comes with a book that identifies which adapter should be used for each manufacturer and the specific engine. All adapters are identified by a number.

The pressure tester with the #21 adapter is used for this 2018 Cadillac ATS with a 2.0L turbocharged engine. The cooling system is holding 15 psi of pressure with no leaks. Most systems should not be pressurized beyond 15 psi. If a greater pressure is used, it may cause the water pump, radiator, heater core, or hoses to fail.

Thermostat Diagnosis

The thermostat is the temperature control device in the engine that sets the engine operating temperature and controls the flow of coolant to the radiator. If a thermostat is stuck in the open position, this will cause the engine to operate too cold. If the thermostat is stuck in the closed position, this will cause the engine to overheat. The thermostat is a controlled restriction.

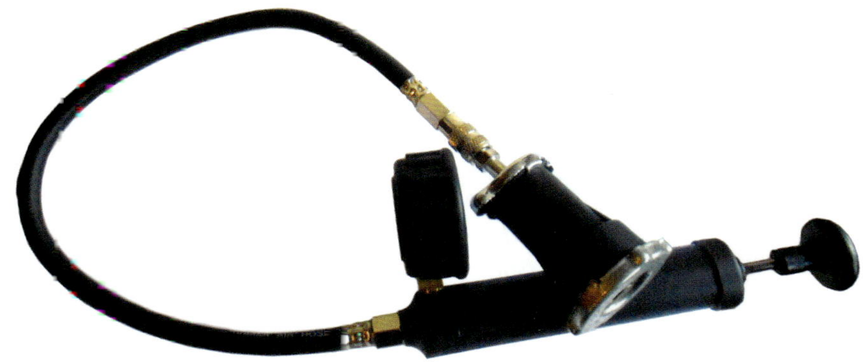

The pressure tester is connected to a specific adapter for testing the cap. The cap being tested is connected to the adapter the same way it connects to the radiator or expansion tank.

Typical Thermostat Ratings		
Temperature Rating	Begins to Open	Fully Open
180°F	130°F	200°F
195°F	195°F	215°F

There are four ways to test a cooling system thermostat for proper operation. The hot water method uses a thermometer in the water, the thermostat, and a feeler gauge. The Tempilstik method uses a calibrated wax stick that melts at a specific temperature to measure thermostat opening. An infrared digital temperature gauge will measure opening when it is pointed at the thermostat housing. Finally, a scan tool will check the engine temperature against the thermostat opening.

Hot Water Method

To diagnose thermostat operation using the hot water method, first remove the thermostat from the engine. Insert a 0.015 inch (0.4 mm) feeler gauge in the opening so that the

The specific pressure listed on the cap is 16 PSI as shown.

thermostat hangs on the feeler gauge. Place the thermostat into a metal pot filled with water. Place a second thermometer in the pot and slowly heat the water on a stove's low setting.

Watch for the thermostat to open enough to release the feeler

gauge. Check the temperature of the water in the pot to determine the thermostat opening temperature. If the temperature is within 5°F of the temperature stamped on the thermostat, it is opening properly.

Tempilstik Method

The thermostat can be tested using a temperature stick called a Tempilstik. It is a pencil-size device that has a wax material containing certain chemicals that melt at a specific temperature. First, look up the opening temperature of the thermostat being testing. For a thermostat rated at an opening temperature of 187°F (87°C), the thermostat valve remains closed below that temperature.

Mark the thermostat housing with a 188°F (87°C) Tempilstik and a 206°F (97°C) Tempilstik. Start the engine and let it heat up to operating temperature. Watch the Tempilstik marks to determine the opening and fully open temperatures of the thermostat. The 188 Tempilstik melts at 188°F (87°C), which is when the thermostat should begin to open. Use an infrared digital temperature

If the thermostat is removed, the coolant could flow more quickly through the radiator. The thermostat adds some restriction to coolant flow. Heat transfer is greater with a greater difference between coolant temperature and air temperature. When coolant flow rate is increased (no thermostat), the temperature difference can be reduced.

A wax pellet thermostat opens when the pellet heats up and then allows coolant to circulate into the cylinder heads in this LT1 Corvette reverse-flow cooling system. To do this test, mark the housing with Tempilstik or use an infrared digital thermometer. (Photo Courtesy Jim Halderman)

gauge to check the temperature of the housing. The 206 Tempilstik melts at 206°F (97°C), which is when the thermostat should be fully open. Again check with the infrared gauge.

If the readings are not between the operating temperature range of your engine, replace the thermostat.

Infrared Thermometer Test

To test the thermostat with an infrared digital temperature gauge, run the engine to operating temperature. Point the digital gauge at the housing and take measurements when the engine is cold, when it reaches operating temperature, and as it cools down. If these readings are within 5 degrees of the temperature for your system, the thermostat is okay.

Testing the hoses around the thermostat with the digital temperature gauge can also tell you if there are blockages in the system or if the fan is not working properly. On a crossflow system, the temperatures should decrease evenly from one side

You can also use an infrared digital temperature gauge to measure the thermostat opening without Tempilstiks. Point the digital gauge at the thermostat housing to measure the temperature when it is cold and as it heats up. This will monitor the temperature when the thermostat is closed, when it opens, and when it is fully open.

to another. On a downflow system, they would decrease from top to bottom.

Scan Tool Method

On vehicles built after 1996, you can attach a scan tool to measure temperature. Attach a factory bidirectional scan tool to the 16 pin data link connector (DLC) under the left side of the dash near the steering wheel. When set on the engine data menu, the tool can read the actual temperature of the coolant as sensed by the ECT sensor, provided the sensor is functioning properly. You can tell this if the check engine light is not on and there are no ECT codes, such as P0116, present or stored.

Thermostat Service

Overheating engines make a knocking sound, which sounds like a ball going back and forth between two metal cabinets. This sound is from a thermostat that is stuck closed, stuck open, or defective. To solve the problem, replace the thermostat.

1. Look up the thermostat replacement procedure in the online service information for your vehicle before proceeding. Your vehicle may be unique and require different steps than other vehicles.
2. Remove the radiator pressure cap and drain the coolant from the radiator drain petcock to lower the coolant level below the thermostat. You do not have to completely drain the system.
3. Remove the hose from the thermostat housing neck and then remove the housing. This allows access to the thermostat.
4. Clean the gasket flanges of the engine and thermostat housing. The gasket surface of the housing must be flat.
5. Place the thermostat in the engine

This engine cutaway shows the place-ment of the thermostat in the thermo-stat housing where the spring chamber and pellet are in the down position. (Photo Courtesy Jim Halderman)

with the sensing pellet toward the engine. Make sure that the thermostat position is correct and in the recessed groove. If not, the housing will tilt when tightened, and this will cause it to leak and the housing can crack.

6. Install the thermostat housing with a new gasket or O-ring.

7. Tighten the housing fasteners to the proper torque and install the upper hose and the correct size of radiator hose clamp. If it is a screw-type hose clamp, tighten it with an inch-pound torque wrench to 22 in-lbs or tighten until the clamp does not move, then add two turns.

8. Fill the system with clean 50-50 mix coolant.

9. Start and run the engine with the radiator cap off to make sure the new thermostat opens and cool-ant is flowing.

10. After determining that the thermostat has opened, install the radiator pressure cap to hand-tight and bleed out any air using the engine bleeder valve

(if equipped). Many engines will self-bleed during the warm-up process when the radiator pres-sure cap is off.

Checking Coolant Protection

The coolant must be tested yearly to determine if it can protect the engine from freezing. Coolant testing ensures that the engine will not suffer from internal corrosion and freeze-ups that could crack the block. Engine coolant contains anti-freeze that keeps the coolant from freezing at low temperatures, and it contains rust inhibitors, which pro-tect the coolant passages that wear out. Coolant can be tested with a hydrometer or a refractometer or by testing the pH level.

Hydrometer

To test the coolant for freeze pro-tection, use a hydrometer to measure the coolant's specific gravity. This determines its density. Draw coolant into the hydrometer and watch the float inside to make sure it rises to a certain level, depending upon the density of the coolant. Antifreeze has a higher specific gravity than water, so the higher the float rises in the liq-uid, the greater the percentage of antifreeze in the mixture.

The typical 50-50 mix of antifreeze and water is demonstrated with this 1/2 cup of water and 1/2 cup of coolant in a container. Place the hydrometer into the cup and draw a sample, trying to keep the bubbles at a minimum. The reading shown is approximately −34°F (−37°C).

Most automotive hydrometers are graduated in temperature in Fahr-enheit and Celsius, going from the lowest to the highest. The more anti-freeze that is present in the solution the higher the float will rise, show-ing the lower the freeze point or the higher the specific gravity. Unfortu-nately, hydrometers are antifreeze specific. So you would need

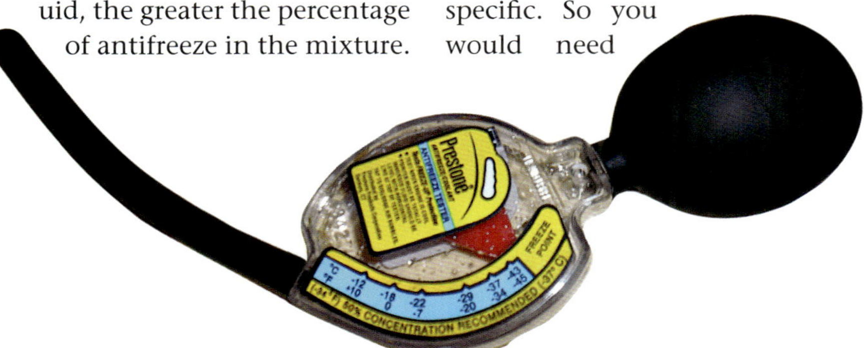

A hydrometer, or aerometer, is an instrument used for measuring the relative density of liquids based on the concept of buoyancy. The higher the density is, the more concentration of coolant in the water. Most coolant hydrometers read the freezing point and boiling point of the coolant.

A container with 2/3 coolant and 1/3 water demonstrates a 70-30 mix. Use the coolant hydrometer to measure and record the freeze protection. Reading is almost off the scale at −43°F (−45°C), but this hydrometer cannot read that higher value.

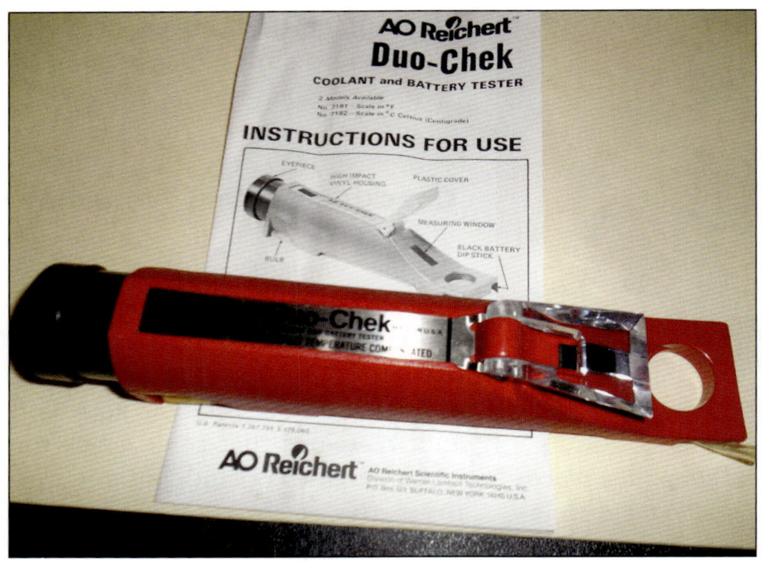

A refractometer is an optical device used to measure the extent to which light is bent, or refracted, when it moves through a substance. It measures the salinity and specific gravity of a liquid.

one for ethylene glycol and one for propylene glycol because they have different specific gravities.

As the temperature of the coolant increases, the specific gravity goes down. Some hydrometers used a built-in thermometer and a chart, allowing you to compensate for the temperature of the coolant.

Refractometer

A refractometer measures a liquid's specific gravity to tell the proportions of antifreeze and water in the coolant mixture. It works by allowing light to shine through the fluid. The light bends in accordance with the particular liquid's specific gravity. The bending of the light displays on a scale inside the tool. The refractometer has a scale for both types of antifreeze.

To use the refractometer, place a few drops of coolant on the sample plate. Hold the refractometer roughly level under a light, look through the viewfinder, and read the scale to verify the freeze protection of the coolant.

Testing pH

The term *pH* comes from a French word meaning "power of hydrogen," and is a measure of acidity or alkalinity of a solution. The pH is based on a scale of acidity from 0 to 14, which tells you how acidic or alkaline a substance is. More acidic solutions have lower pH and more alkaline solutions have higher pH. Substances that aren't acidic or alkaline (neutral solutions) usually have a pH of 7. Test strips that turn color based on the level of acidity are used to test pH.

Testing pH is performed when there is reason to suspect the coolant is contaminated or no longer useful. Testing the pH of coolant determines if the corrosion inhibitors are still working. As corrosion inhibitors break down, the coolant solution becomes more acidic. As the acid

level builds, so does corrosion and electrolysis in the cooling system. If the coolant is left in this acidic condition, it will create permanent erosion to the components. The pH of coolant differs according to the type of coolant used.

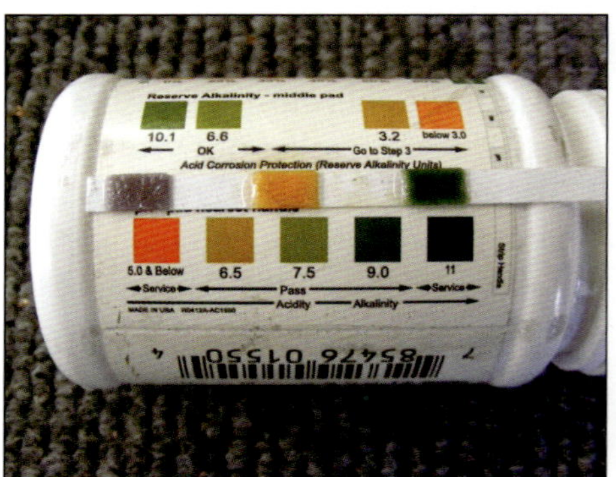

Coolant test strips are used to determine the pH and percentage of antifreeze glycol of the coolant. These strips will change color based on the level of acidity in the coolant and check for freeze point, boiling point, and pH level. The test strips change color when dipped into the coolant, and then you compare the strip to the picture on the container. Always use new test strips from a sealed container. (Photo Courtesy Jim Halderman)

New and Unused Coolant pH Values	
Type	**Value**
Inorganic Additive Technology (IAT)	9 to 10.5
Organic Additive Technology (OAT) or Dex-Cool	7.5 to 8.5 (G30 and G34)*
Hybrid Organic Additive Technology (HOAT)	7.5 to 8.5 (G05, G48, G11, or G12)*
Phosphate Organic Additive Technology (PHOAT)	7.5 to 8.5
*G coolants come from trade name Glysantin of BASF in Europe and Valvoline (Zerex) in the United States.	

A coolant exchange machine helps eliminate the problem of air getting into the system when changing fluids. Air pockets can get trapped in the system and cause overheating or lack of heat. (Photo Courtesy Jim Halderman)

When using a test strip, be sure that it is calibrated to test the type of coolant you have. Used coolant pH readings are usually lower than when the coolant is new. They range between 7.5 and 10 for IAT and lower for used OAT, HOAT, and PHOAT coolants. For best results, an electric pH tester that measures the actual pH of the coolant can be used.

Coolant Flushing

1. Drain system.
2. Fill system with clean water and flushing/cleaning chemical.
3. Start engine; run until it reaches operating temperature with heater on.
4. Drain system and fill with water.
5. Repeat until drain water runs clear.
6. Fill system with 50-50 antifreeze-water mix or premixed coolant.
7. Run engine until it reaches operating temperature with heater on.
8. Adjust coolant level as needed.

Hose Service

Hose service will depend on the type of hose being replaced and the connecting device being used. Many OEMs use spring-type clamps, which can be difficult to remove and install due to tight locations. On a modified system, I recommend using screw-type clamps or the more-robust AN fitting and braided steel or Cool-Flex hoses. If using a screw-type hose clamp, tighten it with an inch-pound torque wrench.

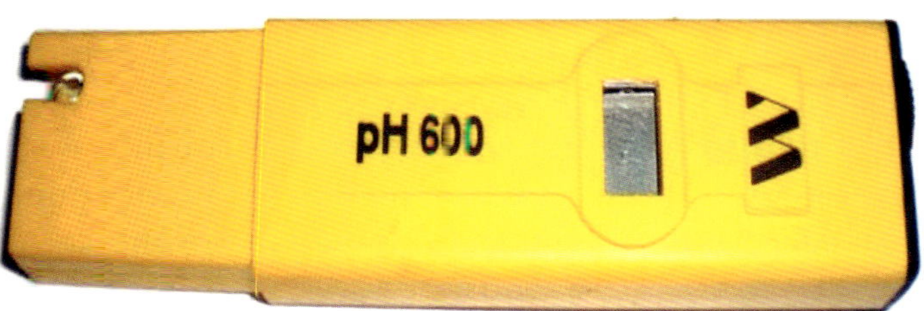

An electronic pH tester measures the actual pH of the coolant and can be used for all coolants, unlike test strips that cannot be used to test the pH of red or orange OAT or HOAT coolants. (Photo Courtesy Jim Halderman)

Spring-Type Hose Clamp Replacement

If a spring-type hose clamp needs to be replaced, also replace the hose as part of the service repair. It is important that a new spring-type hose clamp is orientated in the exact same position as it was installed during vehicle assembly. This ensures that the hose clamp will not be damaged and the hose clamp will not contact or damage any adjacent components. Do not use any adhesive on the hose clamp prior to installation.

When replacing a spring-type hose clamp and hose, the inside and outside hose diameter must be identical between the new hose and the existing hose. This will guarantee proper hose clamp tension and joint integrity. Always place the new hose clamp in the same orientation as it was positioned prior to separating the joint. The new hose might have a marking indicating the correct position of a hose clamp, which should reconfirm the correct hose clamp position. The hose clamp should not contact any adjacent components after installation.

When using the existing hose, always place the new clamp in the

Screw-type cooling system hose clamps are easy to install and tighten as long as the 5/16-inch screw head is in a position to be reached by a 5/16-inch socket and extension. As you can see, they can be located in some pretty tight spaces.

This is a typical OEM spring-type hose clamp, which will lose its tension after years of service. It should always be replaced with a stainless steel screw-type hose clamp.

same orientation as it was positioned prior to separating the joint. The existing hose might have an indentation or compression mark on the hose from where a hose clamp was installed previously. The hose clamp should identically overlay this position. The existing hose might also have a marking indicating the correct position of the hose clamp, which should reconfirm the correct hose clamp position. The hose clamp should not contact any adjacent components after installation.

The water pump being replaced in this photo uses a straight vane impeller with the volute being part of the engine front cover. Most performance high-flow water pumps used on V-8 engines have the volute as part of the water pump assembly. (Photo Courtesy Jim Halderman)

Water Pump Service

This is a simple series of installation steps for the most common types of water pumps driven by a serpentine belt. Always refer to the correct service information for the engine and vehicle you are servicing.

1. Look up the water pump replacement procedure in the service information for your vehicle before proceeding. Your vehicle may be unique and require different steps than other vehicles.
2. Drain the cooling system.
3. Remove the drive belt and accessories in the way. It is very important that you note the way your serpentine drive belt wraps around which accessory pulleys and in what order. There is almost always a belt-routing diagram on the radiator core support showing the serpentine belt routing.
4. Remove the inlet hose from the radiator.
5. Remove the water pump fasteners and remove the water pump. If the water pump is driven by the engine timing belt, you will need to retime the engine when you reassemble the unit.
6. Always discard the water pump seal.
7. Install the water pump into position with a new water pump seal and hand-tighten the bolts.
8. Tighten the bolts to the manufacturer's torque specifications using a torque wrench.

Drive Belt Tension and Replacement

There are four ways OEM-specific belt tension is measured. Look up the drive belt tension procedure in the service information for your vehicle before proceeding. Your vehicle may be unique and require different steps than other vehicles.

Belt Tension Gauge

1. Install belt and operate engine for 5 minutes.
2. Adjust tension to factory specifications.
3. Replace serpentine belt with more than three cracks in a 3-inch span.

Marks on Tensioner

1. Many tensioners have marks indicating normal operating tension range
2. Check service information for locations of tensioner marks.

Torque Wrench Reading

Some manufacturers specify use of beam-type torque wrench to read the tension.

Deflection

1. Depress the belt between the two pulleys that are farthest apart.
2. Deflection should be 1/2 inch or 13 mm.

Always check the service information for the exact marks where the tensioner should be located for proper belt tension. Some belt tensioners will have a mark, in this case a silver arrow, as part of the housing. In this application, it must line up with one of the three marks to the left of the silver arrow. This photo shows the tensioner off of the engine with all tension released.

Here is an example of a typical tensioner as installed on an engine. The tensioner will typically have a square drive hole where you can use a breaker bar or ratchet to release tension.

Radiator Replacement

1. Look up the radiator replacement procedure in the online service information. Your vehicle may be unique and require different steps than other vehicles. You may need to recover the refrigerant from the A/C system, if the condenser needs to be removed to gain access to the radiator. (Photo Courtesy BeCool Performance)

2. Drain the cooling system at the radiator drain valve.
3. Remove the radiator cooling fan assembly.
4. Remove the radiator inlet hose from the radiator.
5. Remove the radiator outlet hose from the radiator.
6. Remove the radiator surge tank inlet hose and the surge tank.
7. Remove the radiator.
8. Replace the radiator by reversing the steps above.

Head Gasket Replacement

The owner of a 1994 Chevrolet Cavalier Sport Touring Sedan with 89,000 miles complained of low heat in the cabin and was continually adding coolant. One evening, the engine severely overheated and was towed to my shop. I installed some water (not coolant) because I did not know the complete nature of the failure. After pressure testing, it was determined there was an external coolant leak at the heater core. In addition, the engine would not start due to being Hydra-Locked. The engine was grunting during the starting process, so a spark plug was removed and water shot out of the plug hole. It was determined that the problem was either a blown head gasket or cracked block or cylinder head.

To replace the head gasket, follow these steps.

1 *When you do any major work on an engine, such as removal of major components (the cylinder head in our case), you should take a digital picture of the engine before removing anything so you know where things go for reassembly.*

2 Disconnect the battery negative cable and then the positive cable. Disconnecting in that order prevents any shorting against the metal body. Remove the battery hold-down and remove the battery.

3 *This is a radiator and cooling system drain valve. Place a pan under this drain valve and open it to drain the coolant. If it will not move, drain the coolant by removing or cutting the lower radiator hose. Discard the coolant according to hazmat disposal procedures.*

4 *It's a good idea to place tags or mark the plug wires with a marking pen. In this repair project, this 2.2L GM 4-cylinder already has the spark plug wires numbered. However, these are the original factory-supplied wires. Aftermarket replacement wires may not have any markings, so make it a practice to always mark the spark plug wires.*

NOTE: All OEMs provide a serpentine belt–routing decal that is usually located on the radiator core support or in a location near the drive belt.

5 Note the serpentine belt routing around the engine pulleys and crankshaft balancer.

Using a 15-mm wrench, rotate the belt tension downward to relieve tension and remove the belt.

6 *Remove the electrical connectors to the alternator and remove the alternator. On this model, there is a top bolt and a bottom bolt. Next, remove the alternator support bracket attached to the head and exhaust manifold. The aluminum support bracket can stay attached to the cylinder head. You do not need to remove it.*

7 *Remove the PCV hose from the intake manifold to the valve cover.*

10 Disconnect the manifold absolute pressure (MAP) sensor green three-pin connector. All OnBoard Diagnostics Generation II (OBD II) input sensor connectors are keyed to their sensors. You do not have to tag or label them.

11 Remove the intake manifold bolts attaching the manifold upper housing to the lower housing. Then, move the intake manifold forward for removal after the vacuum lines are removed.

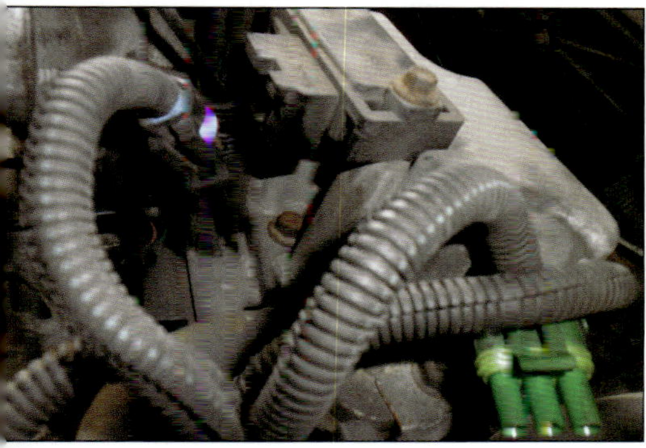

8 Remove the air temperature sensor connector (left). Also remove the air filter housing and air intake tube to the intake manifold (right).

12 Remove the intake manifold vacuum manifold on top of the manifold upper housing by releasing the black plated clasps on each side. Be careful not to break them. Use a small screwdriver to release these clasps. You will also need to remove the vacuum line connection to the HVAC vacuum controls and the charcoal canister solenoid on the back of the engine block. Now the intake manifold upper housing can be removed.

9 Remove the cable cover from the cable bracket, which is attached to the intake manifold. There are two 7-mm screws holding it. Next, remove the accelerator cable, automatic transmission throttle cable, and the cruise control cable. It helps to prop open the throttle valve to remove the cables. Order is not important because they can only go back in the right location. It would still be a good idea to label these cable locations with tags. I am a believer in marking things so they go back in the right place.

13 Remove the valve cover fasteners and the valve cover. Discard the valve cover gasket.

14 Use a 13-mm wrench and/or socket to remove the fasteners of the exhaust manifold. Use a #30 female Torx E-socket or a #30 screw and Vice Grips to remove the exhaust studs. You will not be able to remove this cylinder head unless you remove the studs. You can loosen the exhaust manifold at the header pipe.

16 Look up the head bolt tightening sequence (or order of tightening the head bolts) of your vehicle. Loosen them in that same order a quarter turn at a time until they are loose. If you use an air impact gun and not follow this procedure, you will warp the head and may have to replace the cylinder head.

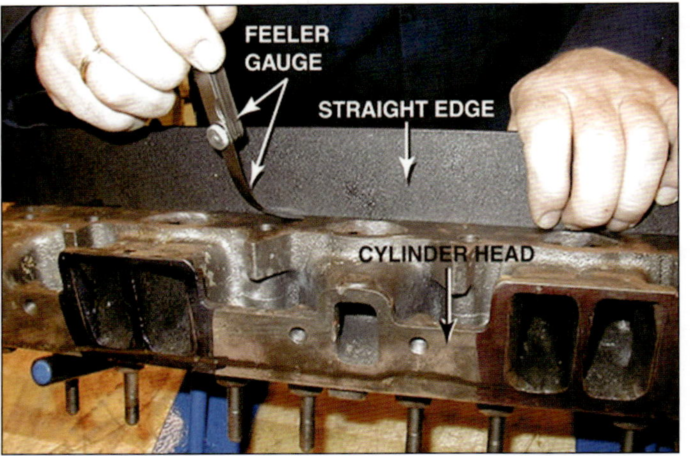

17 Clean the cylinder head and top of the block. Green Scotch-Brite pads in a drill work best, especially on aluminum, which is what this head is made out of. After the head is clean, place a straightedge level on the surface of the head and use the specified feeler gauge to see if it will fit under the straightedge. Cylinder heads should be checked on five planes for warpage, distortion, bend, and twist.

15 Remove the rocker arms and pushrods and place them in a box or container that is marked by exact cylinder number so they can be reinstalled in the same location that they came from. Components will wear in one direction. If that position is changed, the wear may cause a noise. Loosen the rocker arm studs to access the head bolts.

18 Clean all remaining components before beginning reassembly. This particular engine used torque-to-yield bolts that are tightened using the torque-turn method of fastening. In this case, you must replace the cylinder head bolts with new ones, otherwise another head gasket failure will result.

19 *Install the head gasket and cylinder head. The service information contains the correct tightening sequence for the head bolts. The torque tightening sequence starts in the middle and moves outward. Before installing the new head gasket, it is a good idea to spray it with copper gasket sealer to ensure a good seal. Two old head bolts with the heads cut off can be used as guide studs to install the head. New head bolts are installed in the head. Tighten them with a speed wrench (15-mm socket used).*

Tightening Sequence

The tightening sequence is as follows:

a. In sequence, first torque the head bolts to 46 ft-lbs.

b. In sequence, second torque the head bolts to 53 ft-lbs.

c. Tighten each bolt in sequence a quarter turn or 90 degrees.

20 Torque the rocker arm studs to 50 ft-lbs.

21 Install the pushrods and rocker arms in their original position and torque them to 22 ft-lbs. There is no adjustment because they are zero-lash.

22 Drain oil and install a new oil filter and fresh engine oil.

23 Install all remaining components with new gaskets and tighten them to specifications.

24 Fill with fresh antifreeze.

25 Start and run the engine and bleed out all air.

Heater Core Replacement

I am covering the subject of the heater core replacement due to the difficult repair it can be in a number of vehicles. There are many vehicles where this repair is a nightmare. For example, if your vehicle is a GM luxury vehicle from the 1960s or 1970s with a center console, such as a 1965 Buick Riviera, a heater core replacement even for a seasoned technician can take about 8 hours.

I have listed below an example of a heater core replacement on a late-model high-performance vehicle that has 34 specific steps. It is an example of a very involved and time-consuming project that will also require some special equipment. Look up the heater core replacement procedure in the online service information or your factory shop manual for your vehicle before proceeding because there are some steps you might miss that can be detrimental to your vehicle. Your vehicle will most likely be unique and require different steps than other vehicles.

1. Disconnect the negative battery cable.
2. Recover the air-conditioning refrigerant system. If you are obeying state and federal law, you will need an expensive refrigerant recovery and recharging machine.
3. Drain the engine coolant system.
4. Remove the radiator expansion tank.
5. Remove the air-conditioning evaporator thermal expansion valve front tube from the air-conditioning evaporator thermal expansion valve rear tube.
6. Remove the air-conditioning compressor and condenser hose from the air-conditioning evaporator thermal expansion valve rear tube.
7. Remove the engine fuel pump.
8. Remove the fuel pump insulator.
9. Remove the air-conditioning evaporator thermal expansion valve rear tube from the air-conditioning evaporator thermal expansion valve.
10. Remove the heater inlet and outlet pipe from the heater core.
11. Raise the vehicle on a lift or use a floor jack or jack stands and remove the air-conditioning evaporator and blower module drain tube.
12. Lower the vehicle.
13. Remove the instrument panel outer air outlet duct assembly.
14. Remove the instrument panel heads-up display assembly.
15. Remove the left side window defogger outlet duct.

16. Remove the windshield defroster nozzle assembly.

17. Remove the instrument panel tie bar assembly.

18. Temporarily support the heater and air-conditioning evaporator and blower module heater ventilator air-conditioning (HVAC) assembly.

19. From within the engine compartment, reposition the vacuum lines and the underhood wiring harness so it is out of the way.

20. Remove the fuel line clips from the heater and air-conditioning evaporator and blower HVAC module studs.

21. Remove the retaining clip nut and reposition the air-conditioning evaporator thermal expansion valve rear tube and heater inlet and outlet pipe so it is out of the way.

22. Remove the HVAC module assembly right side nut that is holding the HVAC module assembly to the cowl panel.

23. Remove the HVAC module assembly left side nut that is holding HVAC module assembly to the cowl panel.

24. Remove the HVAC module bolts that are holding the HVAC module assembly to the upper instrument panel tie bar.

25. Slide the HVAC module assembly back to clear the air-conditioning evaporator module drain and remove the HVAC module assembly from the vehicle.

26. Remove the right side floor front air outlet duct fasteners.

27. Remove the right side floor front air outlet duct.

28. Remove the left side floor front air outlet duct fasteners.

29. Remove the left side floor front air outlet duct.

30. Remove the heater lower case fasteners.

31. Remove the heater lower case.

32. Remove the heater and air-conditioning evaporator tube seal.

33. Remove the heater core assembly.

34. Install the new heater core and transfer all the necessary components.

Installation would be the reverse of these 34 steps, including charging the air conditioner with the proper equipment. Not all heater core replacements will be this involved, but you can see with these sample steps that this can be quite an undertaking.

Purchase a new heater core from one of the online parts suppliers. Many of them offer OEM parts at a reasonable cost.

Cooling System Design and Service Issues

An efficient operating cooling system needs the right-sized radiator, water pump, thermostat location, good airflow, and cooling fan combination. Your system needs the best water pump speed and coolant flow between the engine and radiator. Generally, when engines overheat or run too cool, it's due to one of the following issues: no thermostat and high coolant velocity, water as coolant, improper coolant levels, missing anti-collapse spring, and the wrong coolant fan is used.

Thermostat and Coolant Velocity

Generally, the thermostat can be located on either the suction or pressure side of the water pump. If on the pressure side, use a 180°F thermostat; on the suction side, use a 160°F thermostat, a 20°F difference. Thermostat selection depends on application. Most rodders choose a 160°F thermostat to treat overheating issues; the 160°F thermostat was originally intended for alcohol antifreeze used before ethylene glycol. The best thermostat for classic vehicle applications today is the 180°F model. If you're having overheating with a 180°F thermostat, you have deeper problems with other components. Late-model computer-controlled vehicles mandate the use of a 195°F thermostat.

Many believe that removing the thermostat helps an overheating issue; others believe this raises engine temperature. When coolant never has a chance to give up heat via the radiator, it gets hotter and hotter. The thermostat adds some restriction to the coolant flow and keeps the coolant in the radiator longer and allows additional time for the heat transfer between the hot engine parts and the coolant. Without the thermostat restriction, much of the coolant flow often bypasses the radiator entirely and returns directly to the engine. Some believe that without a thermostat the coolant can flow too quickly through the radiator.

There is a myth that early Cadillac V-8 engines would overheat if the

This lower heater box cover was removed from the heater core. Note the location of the holes for the fasteners that you might miss when you are under the dash, which is very uncomfortable. If you miss one of the fasteners, you will not be able to remove it.

The heater control box is in the installed position. As you can see, access is limited and difficult, and this application is far from the worst.

thermostat was removed. This has led to the thinking that if the coolant travels to the radiator too fast, it will not have enough residual time to cool off. This myth became firmly engrained in the racing community, so it is common to place a restrictor of the cooling system to slow down the flow. When you slow down the coolant through the use of a restrictor, it does not improve the engine's ability to remove heat. It impedes reducing high-velocity coolant flow.

Coolant will give off more heat when going through the radiator due to higher levels of turbulence. With the amount of restriction in a cooling system and the use of a centrifugal pump, it would be impossible to speed the coolant up to a point that heat dissipation through the radiator would increase the temperature. The faster the coolant moves, the more it will remove heat from the engine and temperature will be much lower. If the restrictor theory were true, then why does the racing industry sell high-flow water pumps when slower would be better. You do not see high-performance cooling system makers offering slow-flow systems. However, thousands of restrictors are sold every year to rodders who also invested in high-flow water pumps.

There is also thinking that there is a net positive change in the liquid cooling temperature when a restrictor is installed. With a coolant restrictor in place, the system pressure builds to such an extent in the area that the boiling point of the liquid is elevated very high. Tests conducted by water pump manufacturers on their water flow test bench have recorded localized pressures as high as 60 psi around the water jacket of the small-block Chevrolet engine with a restrictor at about 6,500 rpm. This action would then raise the boiling point of the liquid in that area. The high boiling point is allowing more heat to be transferred to the coolant and shows a lower temperature reading. This may sound okay, but you are restricting the flow rate for high pressure instead of moving the proper amount of coolant through the system.

I believe in doing whatever works. You could easily try both ways (with and without a thermostat), especially if you are using a remote thermostat. I had a 1965 Buick with a 300 V-8 that had an overheating problem. In my not-knowing-anything years, I decided to try removing the 160°F thermostat to help the overheating by removing the temperature control device. Well, it worked for me; the engine ran cool with no overheating with a marginal reduction in cabin heat.

Water As Coolant

Water is the best coolant in terms of heat conduction, but it is also the best source of corrosion. If you are using straight water, you should always add water pump lubricant and a corrosion inhibitor. You can also use a coolant enhancer such as Water Wetter, which improves surface tension and heat conductivity. Coolant manufacturers often suggest a 50-50 mix of ethylene glycol and water, which will protect your cooling system down to –34°F. There are some that use 100-percent antifreeze with the logic that coolant temperature runs only marginally higher, and this approach eliminates any risk of corrosion.

Today, you can buy antifreeze already mixed with water. If you're going to use an ethylene glycol and water mix, it is also recommended you use distilled water to keep minerals out of your cooling system. Many racing component suppliers sell Evans High Performance Waterless Coolant, which is glycol based.

Improper Coolant Levels

When you are servicing a cold engine, you should add coolant to 1 inch below the filler neck. This allows for expansion as the engine warms. Coolant can rise as much as 1 inch as the engine gets to operating temperature. Always start the engine with the radiator cap removed. As the engine warms, allow time for the thermostat to open and for the engine to remove any air pockets. On some engines there are high spots, where air can be trapped; the engine manufacturer will have an air bleed at this point. During warm-up, open this air bleed to remove the trapped air.

Anti-Collapse Spring Not Used in Replacement Lower Hose

There are some hose manufacturers that do not install an anti-collapse spring in the lower radiator hose. Older engines need an anti-collapse spring in the lower radiator hose because these channel coolant to the water pump and engine under negative pressure, so it can implode or collapse at high RPM. The anti-collapse spring prevents that from happening. Some say it was used only for factory-fill purposes. This has never been true because of the positive pressure on the lower hose during fill. Always use an anti-collapse spring in the lower radiator hose.

Cooling Fans

One belief is the faster a fan turns, the better, which is not completely true. At high vehicle speed, the radiator slipstream should be strong enough to carry heat from the radiator. When air is moving too fast, you get into boundary layer issues where heat doesn't get carried away because air isn't actually touching the fins and tubes. You want air to move slowly enough across fins and tubes to carry heat away. At speeds above 40 mph, your engine doesn't need a cooling fan. This is why a thermostatic viscous clutch fan or electric fan works best.

Some rodders believe more fans are better. Yet, you don't really need a fan both behind and in front of the radiator. Ideally, you will have a fan behind the radiator that sucks the air through and does not impede airflow through the radiator. It should operate based on temperature.

Cooling fans should be shrouded for proper vectoring of air velocity through the radiator. I would look at the factory setup for the engine you are using in your modified vehicle. For example, the Cadillac CTV uses the generation 5 LS Chevrolet Performance LT1 376-ci 6.2L engine with more than 600 hp from the factory. If you are using an LS/LT-series GM crate engine, take a look at the radiator and cooling fan setup.

Your coolant is under pressure to keep the boiling point as high as possible, so you need the highest pressure cap rating suitable for your application. Caps for older vehicles should be rated for 7 to 12 pounds; newer vehicles should have radiator caps rated for 12 to 18 pounds.

When replacing cooling system components such as hoses, water pump, and thermostat, always use the best components. High-performance cooling system hoses last longer than the average off-the-shelf hose, especially when paired with high-quality worm gear clamps. You can find a wide variety of water pumps for nearly every application imaginable. Always choose the high-flow water pump with a crankshaft to water pump pulley ratio of 1:1.4.

INSTALLATION

Now that we have studied all of the cooling system components, we need to bring them together into a performance-based cooling system that will deliver the best engine cooling system that meets the needs of your modified vehicle. I suggest a wholistic approach where the sum is greater that its parts. Consider all of the components that engine coolant touches that carries away the heat of combustion. For example, not just the radiator and water pump but also the thermostat opening, hoses, inlets and outlets, and every turn in coolant flow that might cause it to slow down.

I began this book by telling you there are three dominate areas that affect engine cooling efficiency: radiator surface area, coolant speed, and radiator airflow quantity. These design areas determine the efficiency of the cooling system as expressed in the amount of heat rejection required by the engine. You need to look at the entire system, not just a few components. When you install a large radiator and high-flow water pump, you must also look at the thermostat opening. If you are using a heater core or coolant necks, in other words the path of your coolant, you may

This Plymouth Belvedere modified car shows the installation of a Champion Cooling Systems inline cooling system filter. It will provide an indication of your system's coolant flow and when the thermostat opens. (Photo Courtesy Champion Cooling Systems)

still experience some overheating. You *must* study your entire cooling system.

Gasoline engines do not convert all the energy in the fuel into useful work. Large amounts of the heat generated in the combustion process must be dissipated. The cooling system's purpose is to disperse all excess heat generated by the engine so that the engine runs at its optimum temperature. You should design your cooling system so that the engine does not overheat under maximum loading conditions. Yet, you should not just build a cooling system with a large amount of excess capacity, because this would result in cost and weight penalties. So, one could say that the design of an engine cooling system is a task in optimization. The average cooling system in today's

A typical vehicle coolant temperature gauge shows temperature in degrees Celsius or C (left). There are temperature gauges with the reading in degrees Fahrenheit. (Left Photo Courtesy Jim Halderman; Right Photo Courtesy Bob Wilson)

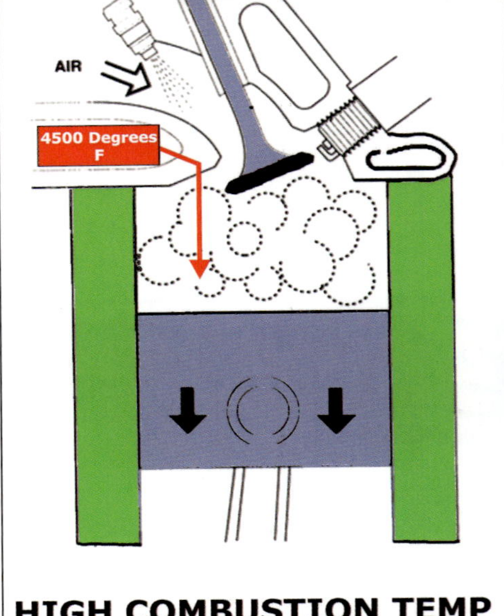

HIGH COMBUSTION TEMP

The temperature in the combustion chamber can get to about 4,500°F The liquid cooling system is in the 212 to 260°F range.

more efficient vehicles weighs about half the weight of the average cooling system in a car from the 1970s, due in part to lighter aluminum components.

When you are converting an original stock vehicle to a modified or race vehicle, you should research the OEM stock vehicle to find out its water pump flow rate and RPM, radiator shrouding, thermostat inner diameter, and coolant system plumbing and routing. I worked for General Motors for more than 25 years, and the company spent a lot of time and money designing an engine to include the cooling system. However, like all OEMs, it must compromise between cost, complexities of mass production, and high performance. I remember the general manager of one of the GM divisions I worked at saying, "The complexities of mass production will never permit perfection." Well, some may say that the Japanese Kaizen (the process of continuous improvement or change to become better) proved him wrong.

After your vehicle is complete and the temperature warning light is not on and the gauge does not read high, your system may not be okay. The typical coolant temperature gauge provides the liquid temperature of the coolant, not the head or combustion chamber temperature. This temperature will more accurately tell you how well your cooling system is working. You cannot easily measure combustion temperature like an OEM, but you can use an infrared thermometer to measure the head temperature.

Start your research using the OEM service information systems. Next, search vehicle forums on the internet for valuable information not found in the service information. Research online publications for high-performance car and trucks for useful performance and building articles, along with tips on your projects. Lastly, go online to the high-performance cooling system component manufacturers such as Champion Cooling Systems, Derale, Meziere, U.S. Radiators, Edelbrock, Holley, and Eastwood for help. Most of these companies have tech tips, forums, online help, and email referrals.

Coolant System Flow Review

Engine coolant is circulated through passages in the engine using a water pump. The pump pulls coolant out of the radiator, where the coolant has given up its heat to the air flowing through the radiator. The lower-temperature coolant will then return to the engine. The radiator can remove the heat efficiently because it contains numerous cooling fins on the air side of the radiator that increase the heat rejection. There is a fan that pulls air though the radiator, increasing the heat rejection.

A control system, which could consist of an electric switch or a viscous clutch, turns the fan on or off. This is done to reduce power loss so the fan is only powered when it is needed. There is also a radiator pressure cap, which pressurizes the cooling system to raise the boiling point temperature of the coolant. The cooling system uses a thermostat to control the flow of coolant through the radiator, allowing the engine to operate near optimum temperature. The engine cooling system will use a coolant expansion or surge tank, where excess coolant is stored to replace any coolant lost from the system during operation.

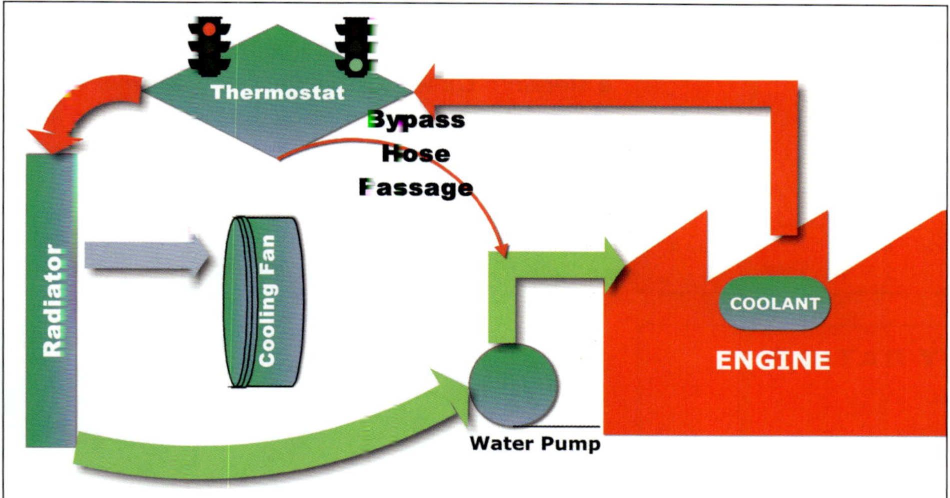

In this coolant flow diagram, low-temperature coolant is being sucked out of the radiator after being relieved of its heat. It then goes into the heads first, then the cylinder water jackets, and then returns through either the bypass hose/passage back into the engine or through an open thermostat depicted as either a green light open or red light closed. It then goes back into the radiator, where the heat will be removed again.

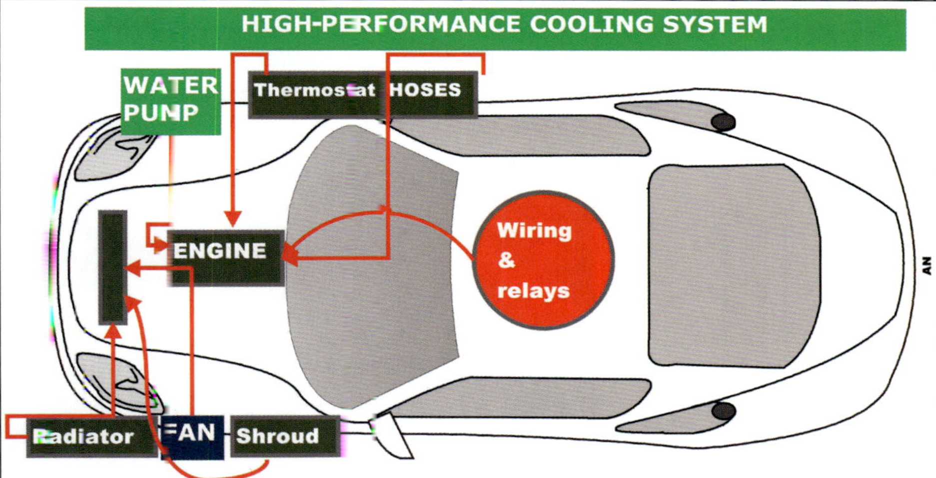

This graphic shows an outline of a modified car and the areas of concern in building your high-performance cooling system.

My approach for the building of a high-flow high-performance cooling system is to start at the engine with the water pump selection, thermostat, and hoses. Then select a radiator, cooling fan(s), and shrouding. This is my suggested order:

1. Vehicle wiring system, if it is designed from the ground up
2. Water pump
3. Wiring for the water pump (if it is electric)
4. Thermostat or no thermostat
5. Hoses and fittings
6. Radiator
7. Shroud

8. Cooling fan (viscous clutch if mechanical fan is chosen; if electric fan is chosen, relays or PWM control)

Wiring Harness

If modifying a stock car or building a race car from scratch, the chassis and suspension components would be built first, just like you build a

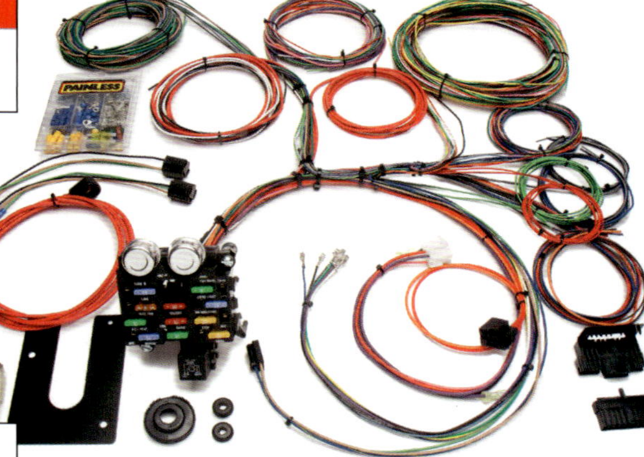

This picture shows a complete vehicle wiring design. It is a Painless Performance Circuit Classic Customizable Chassis Harness (GM Keyed Column 10101). It would most likely be used with a GM engine setup for a LS1 to LS5 or LT1/LT4 engine. (Photo Courtesy Painless Performance Inc.)

house by framing it. The next item that is usually added is the electrical wiring with leads to the projected electrically operated components such as the battery, alternator, cooling fans, electric water pump, etc.

Water Pump Selection

The high-flow pump and outlet with no thermostat combination had the highest GPM flow at 85 gpm. The high-flow water pump with OEM outlet and no thermostat was slightly

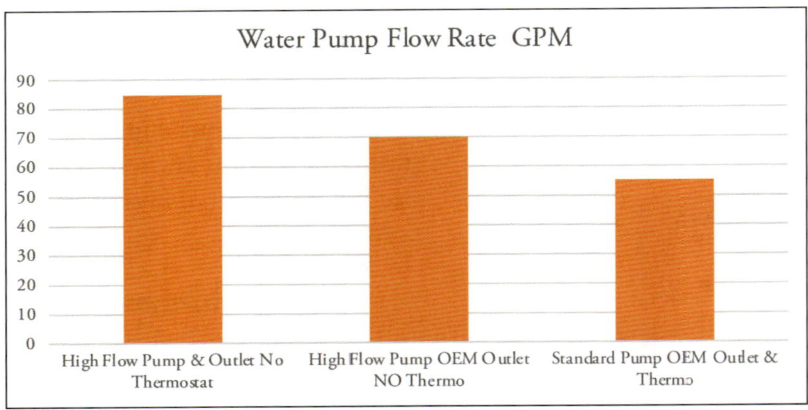

This chart shows a comparison between a high-flow water pump with a high-flow outlet not using a thermostat, the same water pump using a standard OEM outlet to the radiator running no thermostat, and a standard OEM water pump and outlet using a standard thermostat.

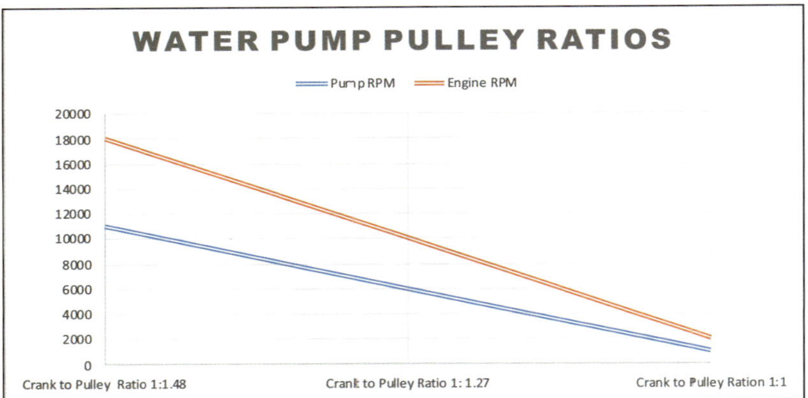

This graph plots the difference in pump and engine RPM for three different water pump pulley ratios. You can see that a crankshaft revolution to pulley revolution of 1 crankshaft turn to 1:1.48 turns results in the highest pump RPM. This will pull the heat out of the system quickly.

The Meziere Enterprises 400-series mechanically driven water pump for a GM LS1 to LS5 V-8 high-performance engine for either production or crate versions is popular for building high-performance street and track vehicles. This pump uses a 4-inch impeller with a high-flow 1.5-inch inlet. (Photo Courtesy Meziere Enterprises)

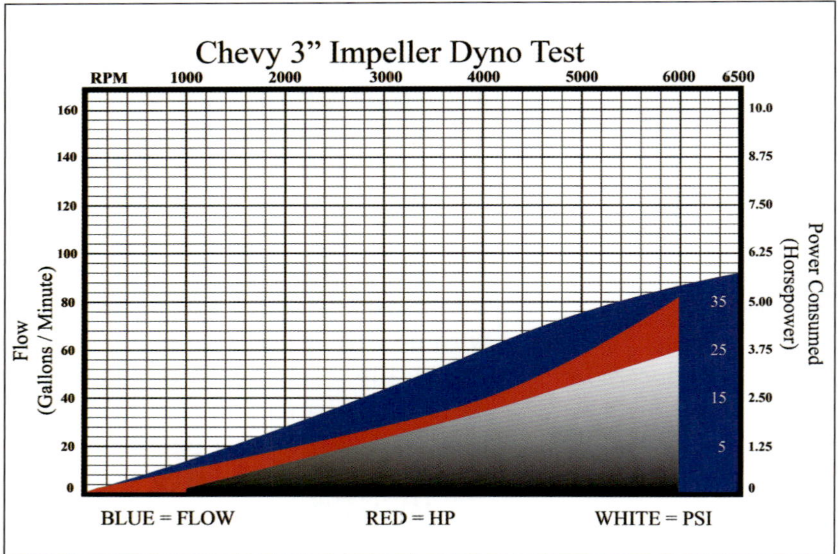

This is a pump performance graph for the Meziere Enterprises electric 300-series water pump with a 3-inch scroll impeller. The graph shows that pump can turn up to 6,500 rpm with a flow rate of 86 gpm, consuming about 6 hp. (Chart Courtesy Meziere Enterprises)

A high-flow water pump is installed on a modified car with a large 1.5-inch coolant inlet from the radiator and an equally large coolant outlet at the thermostat housing return to the radiator.

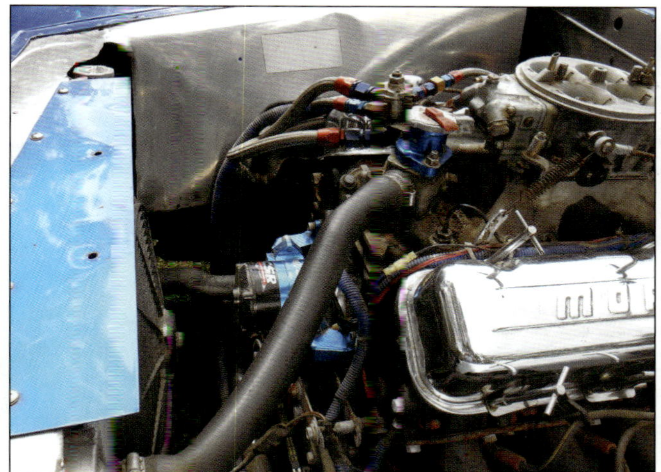

Compare the Meziere 300-series electric water pump with a 3-inch scroll-type impeller installed on a dragster (left) and a Meziere electric water pump installed on a 292 Ford engine (right). (Photos Courtesy Jim Halderman)

lower at a flow rate of 70 gpm. The OEM stock setup using a thermostat was the lowest at 55 gpm. So to get the maximum flow to get rid of the maximum heat, a high-flow pump with no thermostat or a high-flow remote thermostat are needed. Remember your cooling system is only as good as its minimum flow area.

Water Pump Pulley Ratio

When you select a water pump for your project vehicle, you need to consider the following:

- Pulley ratio (should have a pump-to-crankshaft ratio of 1:1.4 or 1:1.5)
- Water pump speed
- Water pump flow rate in GPM
- Electric high-flow pump

I recommend using a high-flow and high-volume water pump such as the Meziere Enterprises 400 series mechanical belt/gear-driven water pump. It is available for the small-block and big-block Mark IV and V Chevrolet engines. This is a superior street and racing water pump that uses a 4-inch scroll-type impeller. This pump will pump RPM up to 6,500 with a flow rate of 120 gpm. It has a high-flow 1.5-inch inlet that goes to the radiator outlet, which is a feature not often seen on performance water pumps.

Electric High-Flow Water Pumps

You may want to consider an electric water pump because sometimes they work better than a stock mechanical pump for street rods. There is another myth out in the modified and racing industry that electric water pumps only worked on drag racing vehicles because the engine cools down between runs by turning on the electric fuel pump.

A Meziere Enterprises high-flow 55-gpm electric water pump was installed on a high-horsepower hot rod, which is also using Flex-Cool aluminum upper radiator hose.

This is a side view of a Meziere Enterprises high-flow 55-gpm electric water pump on a high-horsepower hot rod.

High-flow electric water pumps can roughly provide three times the flow of a mechanical pump at idle. Brushless motor technology has found its way into the electric water pump field like it has in the world of alternators and electric vehicles.

Electric water pumps are generally controlled with a high- and low-speed switch. They can also be wired into your vehicles CAN-BUS system, so the vehicle engine management computer can turn the pump on and off. The typical electric high-performance water pump will deliver about 55 gpm.

Water Pump Installation

The installation of a replacement water pump is pretty straightforward. Your engine will most likely not have the encumbrance of all the hang-on devices and accessories found on a new-production engine. Not so with the installation of a 1992 to 1996 GM Chevrolet LT1 5.7L V-8 engine water pump. If you are using this engine and replacing the pump

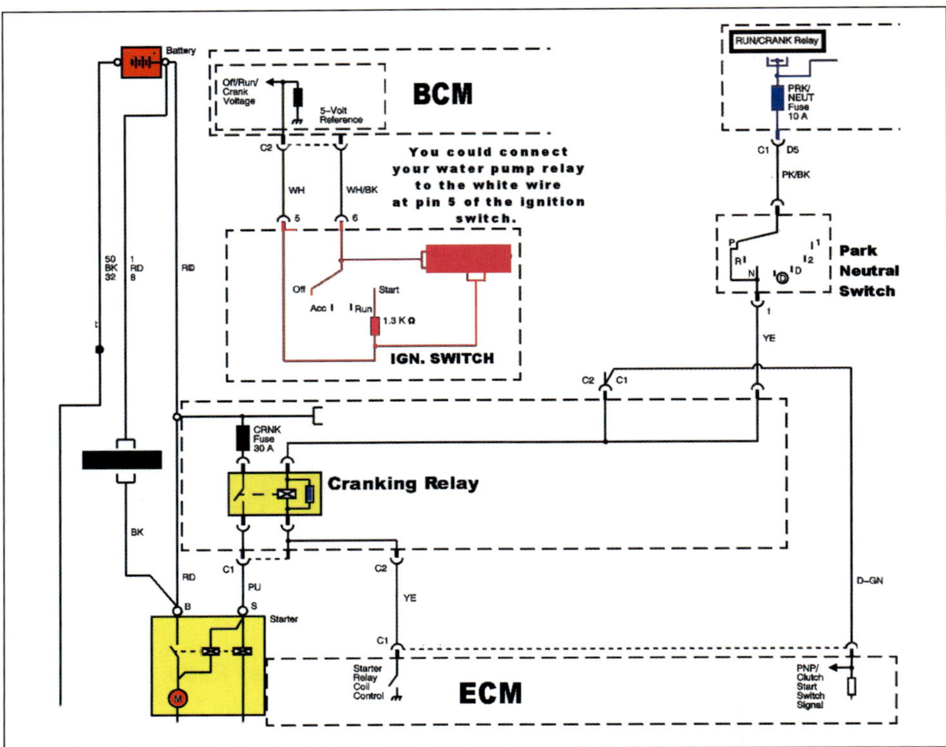

This sample wiring diagram shows some possibilities to hook up an electric water pump to be controlled by the action of the body computer module (BCM).

with an OEM or aftermarket pump, the installation is a little different because the water pump is driven by a small splined shaft from an idler gear driven by the camshaft to the water pump.

Aftermarket mechanical and electric high-flow water pumps are designed to fit on the engines where the OEM pump was installed with little modification. You have some minor changes to make to the brackets and hoses. However, if you are using a Meziere electric high-flow water pump, you will need to add the wiring to run the pump. Many

Shown here is a modified vehicle that the builder has installed a 1994 Chevrolet Corvette LT1 engine. It uses a reverse-flow cooling system with the camshaft-driven water pump.

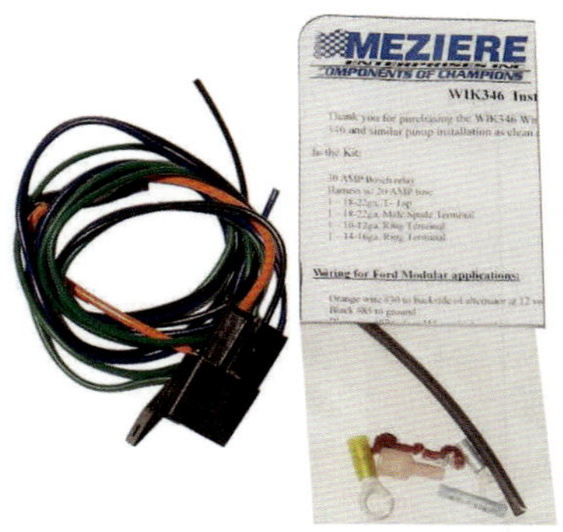

A Meziere water pump relay kit can be purchased separately or is part of the water pump installation package that comes with the new electric water pump. (Photo Courtesy Meziere Enterprises)

NHRA Pro Stock

Electric water pumps are the choice of NHRA Pro Stock champions Greg Anderson and Jason Line to keep cool in the heat of battle. The Meziere 300-series pumps changed the rules about using electric pumps on high-horsepower street engines, nitrous motors, or super/turbocharged cars. Delivering 55 gpm, the 300-series pumps offer great cooling solutions to high-horsepower vehicles. Higher flow rates reduce the chance of detonation. ■

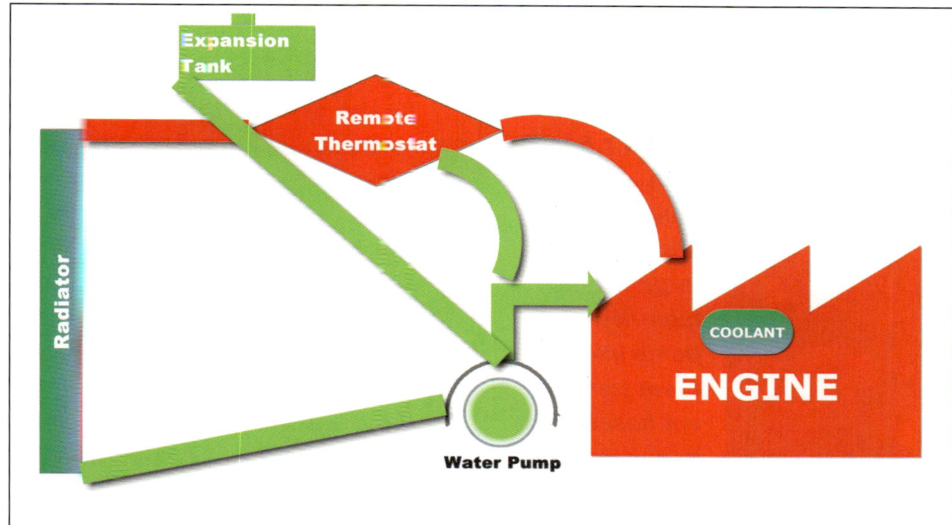

The engine cooling system thermostat is in a remote location, coming from the hot engine to the remote thermostat to the radiator inlet. There is also a hose plumbed from the remote thermostat to a new inlet as a bypass for a cold engine connected to the water pump inlet. This is also the path for coolant from and to the expansion tank.

of these pumps come with an installation kit that includes wiring and a relay. Meziere also markets a relay kit for use in installing the electric water pump.

If modifying a late-model vehicle that uses network communication such as a CAN-BUS or GM Local Area Network (GMLAN), connect the electric water pump into the network so the engine management computer can operate the water pump based on input sensor data. This can be done for the electric cooling fans as well.

You will need to go the online service information and download the wiring diagram for the CAN-BUS and engine management computer for your specific application. Remember, you must provide circuit protection in this circuit by using a fuse or fusible link and connect the pump circuitry to an adequate voltage supply. If you are having difficulty knowing where to connect the pump even with the diagrams, you can contact Meziere and its cus-tomer service can help you. You can connect a Meziere relay to a number of low-voltage connections to operate the water pump.

Thermostat, Inlet, and Outlet Selection

The flow of coolant through your cooling system must not be overlooked. Installing a big radiator with multiple passes may not be enough to remove the engine heat if you have a slow water pump and restrictions in your coolant flow.

The thermostat becomes a key cooling system component for increasing engine warm-up and heater core performance if you are using one. Yet, it can be a very large restriction in your coolant flow even when fully open. If you are going to use one, I suggest a remote thermostat due to its higher flow rate.

Traditionally, the thermostat was located in the intake manifold, where its fluid opening was limited. Meziere makes a series of larger high-flow remote thermostat housings and thermostats. You can also use a larger outlet and inlets to the radiator for better coolant flow to reduce any chance of overheating.

The thermostat selection or the lack of one depends on the vehicle use and the chemical composition of the coolant. If you are just running water, you will need a thermostat with a lower rating due to the low boiling point of water at 212°F. The same is true if you are using the tradition 50-50 mix of ethylene

Champion Cooling Products has an aftermarket thermostat housing that is mounted on top of the original thermostat, opening in the intake manifold. Its purpose is a higher flow rate than a production thermostat setup. (Photo Courtesy Champion Cooling Systems)

This cooling system installation uses a 1.5-inch, high-strength polymer radiator hose on a large 1.5-inch high-flow water pump inlet. Clear transparent hoses help to ensure that the coolant is flowing in the system. You can also see the AN fitting installed on a rubber line stainless steel braided hose to the expansion tank.

glycol and water. If you are using a high-performance coolant such as Evans waterless coolant, be mindful that at temperatures toward –40°F Evans coolant may become thicker. You will need a higher-rated thermostat to keep the cylinder head surface temperature intact.

Coolant Hoses

Cooling system plumbing includes: the upper and lower radiator hoses, heater core hoses (commonly known as heater hose in the 5/8- and 3/4-inch inner diameter), and the inlets and outlets from the engine and other cooling system components. They will carry the hot and cold coolant to and from the radiator.

There many different styles of hoses and fittings, ranging from simple gear clamps on rubber hoses to nylon-lined stainless steel braided hoses with AN fittings. Many builders of high-performance cooling systems use the AN fittings connected to stainless steel braided hose due to their strength. They offer similar performance but each fitting system has its advantages and drawbacks. When selecting a hose design, consider availability, appearance, performance, cost, ease of service, required tools, and ease of installation.

On a race car, serviceability is paramount, but on modified vehicle, appearance is king. The needs of your project vehicle and its engine will dictate the size and type of plumbing that you use. You need to calculate every twist and turn in your plumbing to make sure there is nothing in your system that will impede the flow of coolant to carry away the heat of combustion.

Cooling System Hose Choices

As the vehicle builder, you have to consider durability, pressure capacity, chemical compatibility, aesthetics, tool requirements, and availability of the cooling system hose.

Reinforced Rubber (Low-Pressure) Hoses: Good enough for most applications; readily available; for specific applications (not universal).

Universal Accordion-Style Rubber Hoses: Rigid and can cause cracks in radiator inlets and outlets.

Stainless Steel Braided Nylon-Lined Hoses: Can withstand high pressure; inside is nylon lined and outside braiding is electrically conductive; may or may not be DOT approved; reusable fittings; some may be coated or wrapped on the exterior with Kevlar, which can be found in some automotive applications; outer layers can add strength, prevent dirt and debris from getting between the braiding and the rubber, and aid in assembly by covering the wires when placing fittings over the line; expensive.

Stainless Steel Braided Rubber Hose: Low- to medium-pressure systems, such as the cooling system; use with AN fittings.

Cool-Flex Hoses: Flexible aluminum that can be molded into the desired shape.

Hose and Clamp

The rubber hose and gear-type clamp is the most common connection system used for cooling systems. High-quality pre-formed radiator hoses for specific applications are preferred over universal accordion-flex hoses. Universal flex hoses and braided stainless steel hoses are relatively stiff and can place a load on the radiator neck, which can cause cracking, so support mechanisms must be used.

Cooling System Fitting Types

- Hose and screw clamp systems
- AN/JIS 37-degree flare fittings
- AN fitting installed in a rubber/nylon-lined hose
- AN O-ring seal (ORS) fittings
- Spring-lock clamps

A modified car where all of the cooling system plumbing has been accomplished.

Radiator core support mounting pads can be purchased from radiator firms. (Photo Courtesy BeCool)

Radiator Selection

A 2.5-inch gear-type screw hose clamp is used with molded radiator hoses and flexible accordion-type hoses (top), and an OEM spring-type clamp that should be discarded and replaced with the gear-type hose clamp (right) is shown here.

Rubber hose must have woven fabric reinforcement. It must not collapse under low pressure; a spring is installed in the lower radiator for this purpose.

Cool-Flex hoses (polished aluminum flex hoses) are an acceptable and attractive "flex" solution for street machines when the hose is a moderate length. Once these are bent to the desired shape, they retain their shape and don't stress the radiator like the typical all-rubber accordion hose does.

Gear-type clamps should be retightened when the system is hot, but do not overtighten them. The water pump inlet and radiator or expansion tanks tubes should have raised beads to improve sealing and hose retention. Bead-rolling tools are available to place a beaded end on thin wall tubing. Beads can also be welded onto tubes.

After the water pump, radiator selection is the next most important area of your cooling system installation. Before you design what type of radiator (OEM, brass construction, aluminum, etc.), you need to measure the space that the radiator will occupy.

Measure the width, height, and depth so you can select the most-appropriate water pump. If you have an existing radiator core support, measure it; if not, measure the vehicle space and purchase a support that fits your radiator, modify that OEM support, or fabricate an entire radiator core support. Most aftermarket and OEM radiator manufacturers offer help in this regard, as well as rubber core support mounting pads.

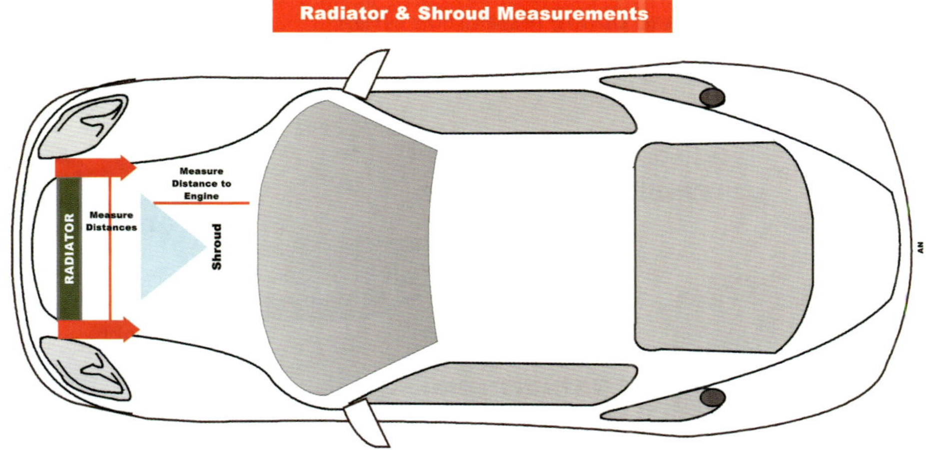

You will need to measure the space that your radiator and shroud will occupy. This illustration provides a rough idea of measurement and space for the radiator and shroud.

Installing AN Thread Fitting

A #20 AN fitting can be installed on OEM-style traditional rubber-molded hoses. Note: This process can be done on a stainless steel braided rubber line hose or a pre-formed rubber OEM hose. AN fittings use a 37-degree flare.

Shown here is a typical #20 AN flare-type end fitting on a braided stainless steel nylon-lined hose for a flexible upper radiator hose.

A #20 AN fitting is used during installation on an OEM-style traditional rubber-molded hose on this modified Ford.

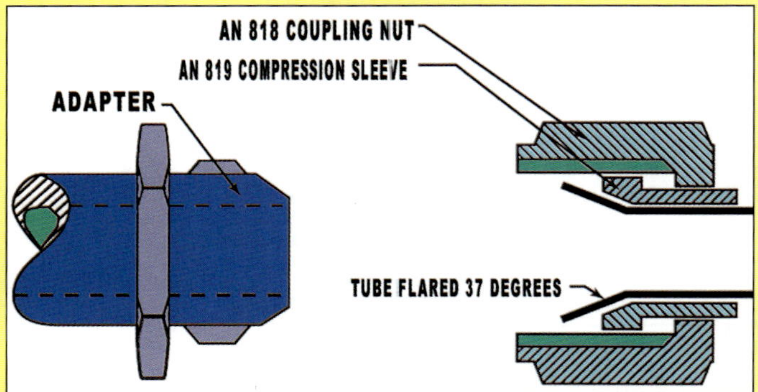

The AN fitting 37-degree flare should never be mixed with the standard 45-degree line flare.

1

Cut the hose for #8 AN fitting installation. Hoses are commonly cut with a cut-off wheel on a grinder, large electrical cable cutters, or a sharp chisel and hammer. Clamp the hose in a hose-clamping device like the one shown or an aluminum or brass soft-jawed vice. Wrap some painters tape or duct tape at the point of the cut before cutting the hose using one of the mentioned methods. (Photo Courtesy kangamotorsports.com)

A cut-off wheel works great but it generates debris that can get into the hose. (Photo Courtesy kangamotorsports.com)

There will be debris in the hose after when using a cutting wheel. Clean out the little bits of burnt up hose, otherwise it will end up in the radiator or engine. (Photo Courtesy kangamotorsports.com)

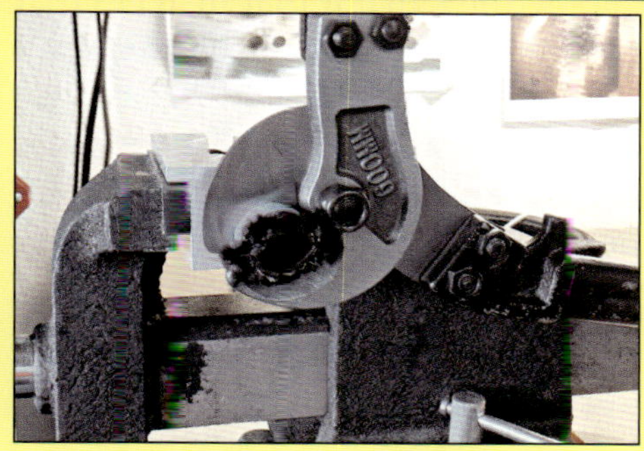

Electrical cable cutters cut through the hose with sharp cutting teeth. Use masking tape to prevent the braid from fraying. The problem with cutters is that it can crush the inner lining hose. You might need to reshape it to round with a drift. Another option is to use a sharp chisel for a clean cut, though it can take a few hits with the hammer to get through. Using a chisel won't produce debris and when combined with masking tape can prevent fraying. Similar to electrical cutters, this can also crush the inner hose lining and gets worse with a lot of hammer hits. (Photo Courtesy kangamotorsports.com)

2

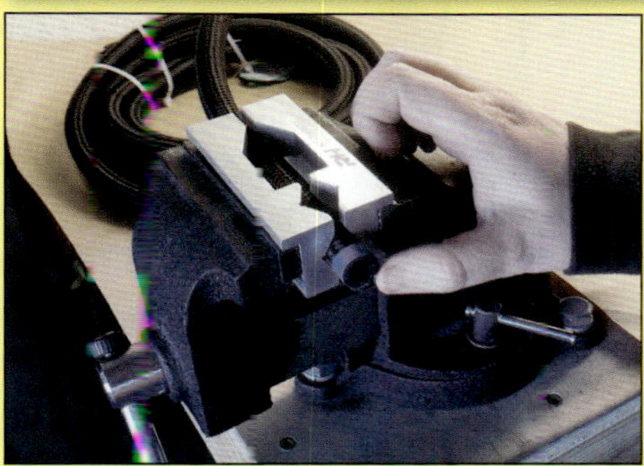

With the hose still clamped in the bench vise, clean up the ends. Depending on the cutting method, you may have some uneven ends. If you used the cut-off wheel, make sure to scrub out the debris from the hose. Compressed air won't get the little bits of burnt up hose. If you used the other methods, you might need to reshape the inner liner to round with a drift or a Phillips screwdriver. (Photo Courtesy kangamotorsports.com)

3

Use a clamping fixture or a soft-jawed vice. Clamp the female AN fitting in the fixture or vice and push the hose down into the socket. Twist as you insert it until it bottoms out. (Photo Courtesy kangamotorsports.com)

4

Clamp the male end of the fitting into the fixture or soft-jawed vice. Apply a lubricant to the threads and start hand-threading the socket down onto the fitting. Be careful not to strip the threads when starting. (Photo Courtesy kangamotorsports.com)

5

Once the hose goes over nipple, tighten using either an assembly tool or a standard AN wrench. (Photo Courtesy kangamotor sports.com)

Service note: Mixing AN-type fittings with another type can cause leakage at the flare. ■

This photo shows a completed rubber-lined stainless steel braided upper radiator hose with a #20 AN fitting.

Once you have secured the radiator measurements, shop for the best radiator that fits your needs and budget. Virtually all of companies have helplines and email assistance in determining the best radiator for your project vehicle. There will be photographs of potential radiators and installation in this section.

Limited Core Support Width

Your modified vehicle project may be an older vehicle from the 1930s or 1940s, where you have limited space for a radiator and you cannot use a large crossflow design. Most of the aftermarket radiator companies, such as Champion Cooling and BeCool, offer radiators that will work in a vehicle with this limited space.

Large Core Width

If you are modifying a newer muscle car, light truck, or standard-size vehicle, you will most likely have enough core width to support a large crossflow radiator of single-, double-, or triple-pass flow design. In this case, the selection from all the radiator companies is vast. If the modifications are not extensive, you will most likely be using the original core. When you choose the radiator that matches that core, installation will be quite easy.

The radiator should be as thin as possible and have the least number of fin density (about 0.7-inch deep single-row radiator for your first choice) to establish a performance curve. Let's say you are selecting a radiator for a high-horsepower crate engine to be installed in a modified street vehicle. Consider a dual- or triple-pass radiator. During your installation, try to find service manual diagrams and installation directions on a vehicle of your type to help your radiator installation.

Large crossflow radiators with a triple-pass flow are complete with dual fans and shrouding ready to connect to your cooling fan control system. This setup is priced at more than $1,400. (Photo Courtesy BeCool)

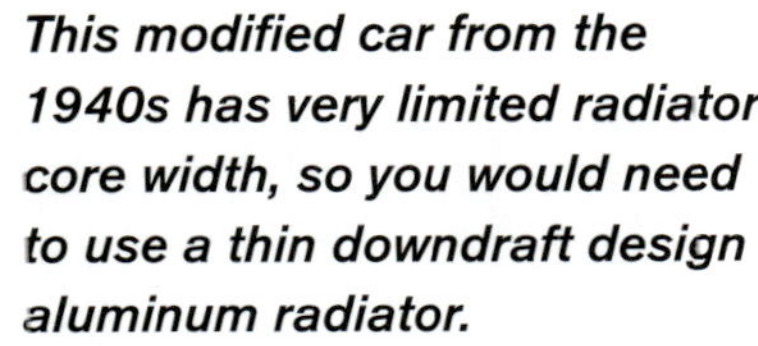

This modified car from the 1940s has very limited radiator core width, so you would need to use a thin downdraft design aluminum radiator.

BeCool offers a downdraft radiator for vehicles with limited core width. This radiator has an automatic transmission cooler in the bottom tank, so it is intended for an automatic transmission vehicle. They do make the same radiator without the transmission cooler. (Photo Courtesy BeCool)

A racing-quality BeCool complete radiator and fan kit with wiring and an expansion or surge tank is available. This unit is ready to go as long as it fits your core support. (Photo Courtesy BeCool)

This radiator is in need of replacement. It also shows the typical radiator core support installation.

You can install a radiator complete with cooling fan, or you can install a bare radiator and add the electric cooling fan. A mechanical cooling fan would already be in place.

Radiator Positioning

The position and placement of the radiator is critical to cooling system performance. You should place it in an area that has good airflow. For some designs, this may not be possible. It really depends on the size and placement of the radiator core support.

The radiator may be very low in relation to the engine, such as in an NHRA Fuel Dragster or Funny Car. If this is the case with your project car, you must install an expansion tank using a pressure cap at the highest point possible. This is necessary because if the radiator is located at the lowest point in the coolant flow path, it will be very hard to fill the system and bleed out the air. Some dragster builders mount the radiator on top of the vehicle behind the engine and use an electric fan to push air through the radiator.

Here is another tricky radiator position on a vintage hot rod.

This mounting setup on a fuel dragster has the radiator set in low position in relation to the engine. All installations are different. The installation should be made with the maximum airflow through the radiator. (Photo Courtesy Champion Cooling Systems)

This builder went to extraordinary lengths to ensure that all the air entering the grille would travel through the radiator. Building small-block off-plates between the radiator and the core support to direct all the air through the radiator will accomplish the same thing.

This setup shows a modified car with a Chevrolet 383 stroker using an aluminum downflow radiator with an overflow tank to the left. There is an aluminum air-conditioner condenser in front of the radiator that should be included in your radiator airflow calculation.

Due to limited space in the radiator core support, this setup uses an aluminum downflow radiator that is not very wide due to a 1930s vehicle design.

This setup in a modified Chevrolet uses a traditional brass downflow radiator with a modified shroud and electric cooling fan.

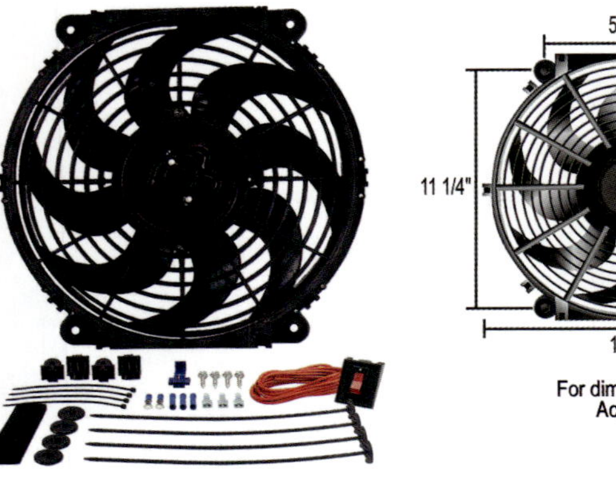

A Derale Performance Model 16512 electric cooling fan with mounting kit and wiring is shown (left). The dimensions for installing this fan kit must also match the radiator dimensions, and you will need a mounting kit to install the fan to your radiator (right).

Electric Cooling Fans

A mechanical cooling fan would have been part of your engine build. Mechanical cooling fans use rigid curved blades that are noisy and absorb a measurable amount of engine horsepower at higher speeds. Instead, you can use a flex-fan design to increase fuel economy, decrease noise, and provide more airflow at idle and flatten out at higher speeds.

When selecting an electric cooling fan, make sure that what you are using matches the dimensions of your new radiator. There is a variety of electric cooling fans to choose from, and the makers of these offer them in sizes that match most radiator sizes.

If you are using a PWM modulated fan controller, there are packages available for specific fans to mount that controller, or you can purchase a PWM controller kit and install it yourself. The high-output Derale fan comes with an installa-

Here is a single electric fan installation using a scroll fan blade for maximum airflow.

Derale Performance makes a fitted fan that goes directly on the radiator. It matches the dimensions shown on the right of this picture. Derale does not make radiators, but the fan sizes match most all radiators. (Photo Courtesy Derale Performance)

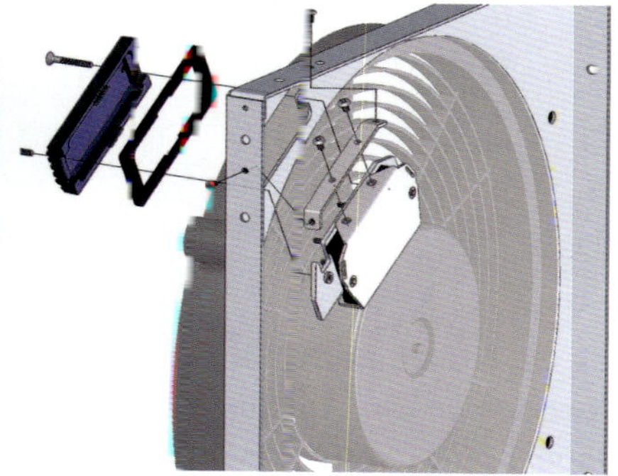

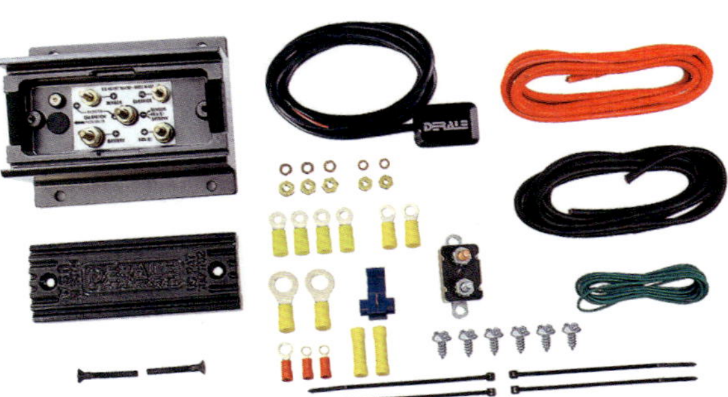

A separate PWM controller installation kit can be mounted in a suitable location in the insert opening provided on some cooling fan assemblies. The kit contains a temperature sensor, circuit breaker, attaching hardware, wiring, and specific instructions. (Photo Courtesy Derale Performance)

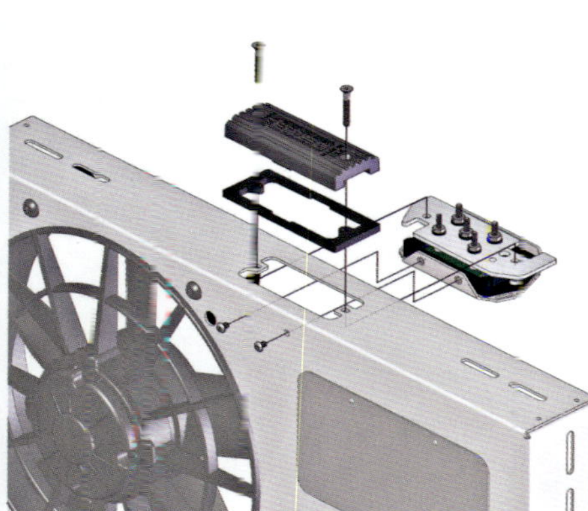

Derale fitted cooling fan setups have cutout areas in the fitted fan assembly where you can install the Derale PWN electric cooling fan controller in several different locations (top is side mount, bottom is top mount). (Photo Courtesy Derale Performance)

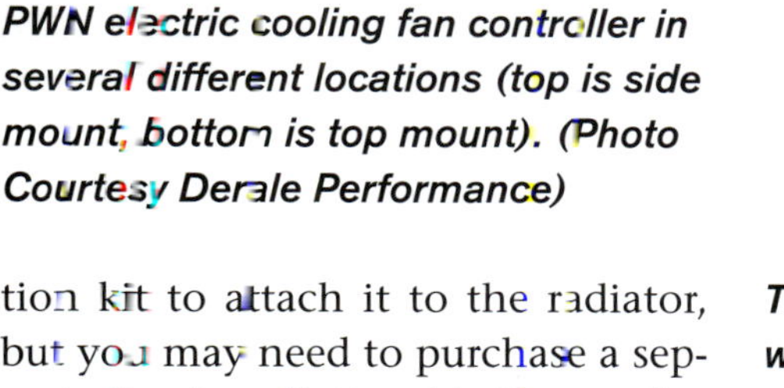

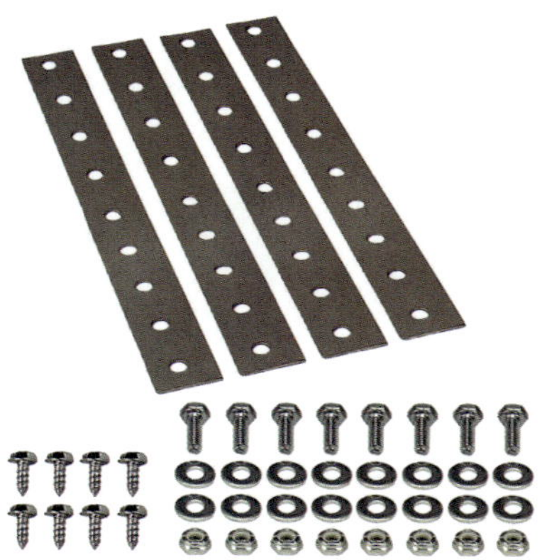

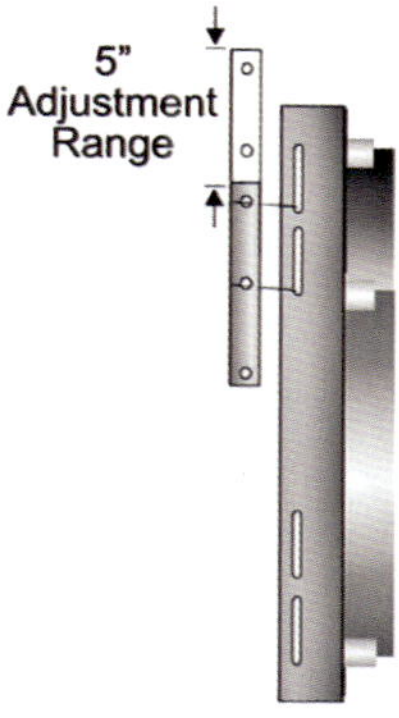

tion kit to attach it to the radiator, but you may need to purchase a separate fan installation kit if your fan or radiator does not come with one.

There are separate cooling fan installation kits for a fan that does not come with one. The flat steel strips with holes in them attach to the side of the cooling fan housing vertically (left). Then can be used to attach to the radiator (right). Fastening hardware is included. (Photo Courtesy Derale Performance)

Expansion and Overflow Tanks

If you are following an OEM design, your cooling system is ideal if it uses an expansion or overflow tank with a pressure cap. The expansion tank should be located at the highest point in your cooling system.

Coolant

Auto manufacturers spend millions on research to make their engines last as long as possible, so take a tip from them on what coolant to use in your modified vehicle based on which OEM engine you are using.

Evans Heavy Duty waterless engine coolant is recommended for racing engines run on tracks or in series where there is no ethylene glycol rule. It is also authorized for use in NHRA racing venues. Evans makes several different coolants; High Performance, Powersports, Heavy Duty and NPG (non-propylene glycol).

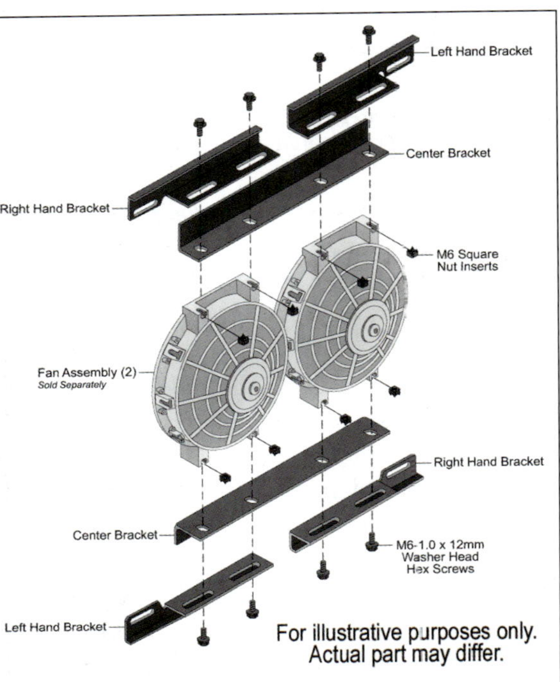

Another type of installation kit is offered by Derale to install a dual cooling fan set to the radiator. (Photo Courtesy Derale Performance)

An expansion tank installed on a 427 Ford Cobra (left) or a scroll-blade electric cooling fan and aluminum radiator (right) can be used in your cooling system design.

A modified vehicle with a fuel-injected engine is using an expansion tank with a radiator pressure cap shown at the right side of the engine.

This OEM radiator uses a dual electric cooling fan installation on a an engine with no expansion, recovery, surge, or catch tank. The hose coming out of the bottom of the radiator pressure cap drains to the ground.